24th International Conference on Computational Statistics (COMPSTAT 2022) and CSDA & EcoSta Workshop on Statistical Data Science (SDS 2022)

Bologna, Italy
23 – 26 August 2022
26 – 28 August 2022

ISBN: 978-1-7138-7017-3

Printed from e-media with permission by:

Curran Associates, Inc.
57 Morehouse Lane
Red Hook, NY 12571

CURRAN ASSOCIATES INC. proceedings.com

Some format issues inherent in the e-media version may also appear in this print version.

International Statistical Institute (ISI)
P.O. Box 24070
2490 AB The Hague
The Netherlands

Phone: 31-70-3375737
Fax: 31-70-3860025

www.isi-web.org

Additional copies of this publication are available from:

Curran Associates, Inc.
57 Morehouse Lane
Red Hook, NY 12571 USA
Phone: 845-758-0400
Fax: 845-758-2633
Email: curran@proceedings.com
Web: www.proceedings.com

PROGRAMME AND ABSTRACTS

24th International Conference on

Computational Statistics (COMPSTAT 2022)

http://www.compstat2022.org

Plesso Belmeloro, University of Bologna, Italy
23-26 August 2022

CSDA & EcoSta Workshop on

Statistical Data Science (SDS 2022)

http://www.compstat2022.org/SatelliteWorkshop.php

Department of Statistical Sciences "Paolo Fortunati", University of Bologna, Italy
26-28 August 2022

COMPSTAT 2022 Scientific Program Committee:

Ex-officio:

COMPSTAT 2022 organiser and chairperson of the SPC: Alessandra Luati and Maria Brigida Ferraro.
Past COMPSTAT organiser: Cristian Gatu.
Next COMPSTAT organiser: Erricos Kontoghiorghes.
Incoming IASC-ERS Chairman: Cristian Gatu.

Members:

Peter Filzmoser, Christian Hennig, Tsung-I Lin, Martina Mittlboeck, Domingo Morales and Miguel de Carvalho.

Consultative Members:

Representative of the IFCS: Berthold Lausen.
Representative of the ARS of IASC: Philip Yu.
Representative of the LARS of IASC: David Fernando Munoz Negron.
Representative of CMStatistics: Ana Colubi.

Local Organizing Committee:

Alessandra Amendola, Enea Bongiorno, Fabrizio Durante, Marzia Freo and Paolo Giordani.

SDS 2022 Scientific Program Committee:

Members:

Elvezio Ronchetti, Ivan Kojadinovic, Bertrand Clarke, Xinyuan Song, Michele Guindani, Peter Winker, Chenlei Leng, Stefano Castruccio, Taps Maiti, Igor Pruenster, Hans-Georg Mueller, Juan Romo, Cheng Yong Tang and Jane-Ling Wang.

Organizers:

Ana Colubi, Erricos Kontoghiorghes, M. Brigida Ferraro, Marzia Freo and Alessandra Luati.

Dear Colleagues and Friends,

We wish to warmly welcome you to Bologna for the 24th International Conference on Computational Statistics (COMPSTAT 2022) and the CSDA & EcoSta Workshop on Statistical Data Science (SDS 2022). After two years of postponements due to the pandemic, we are especially grateful to all those who have kept re-organizing their plans and agendas to join us for these events, either in person or virtually. For many of us, this will be the first opportunity to network in person since 2019. It will be, thus, a special occasion, and we have endeavoured to make it memorable.

These events are locally organized mainly by members of the University of Bologna and The Sapienza University of Rome, assisted by renowned international researchers. The COMPSTAT is an initiative of the European Regional Section of the International Association for Statistical Computing (IASC-ERS), a society of the International Statistical Institute (ISI). COMPSTAT is one of the most prestigious world conferences in Computational Statistics, regularly attracting hundreds of researchers and practitioners.

The first COMPSTAT conference took place in Vienna in 1974, and the last edition took place in Iasi, Romania, in 2018. It has gained a reputation as an ideal forum for presenting top-quality theoretical and applied work, promoting interdisciplinary research and establishing contacts amongst researchers with common interests.

Keynote lectures are addressed by Prof. Igor Pruenster, Bocconi University, Italy, Prof. Jean-Michel Zakoian, ENSAE, France, and Prof. Holger Dette, Ruhr-University of Bochum, Germany.

More than 550 submissions have been received for COMPSTAT, and about 450 have been retained for presentation at the conference. The conference programme has 50 contributed sessions, 8 invited sessions, 3 keynote talks, 54 organized sessions and 2 tutorials. There are approximately 520 participants. For the first time, the conference will be hybrid, and all the sessions will be live-streamed so that participants can attend online the full conference.

The CSDA & EcoSta Workshop on Statistical Data Science has about 65 participants and 50 talks. SDS keynote lectures are addressed by Prof. Geoffrey McLachlan, University of Queensland, Australia, Prof. Peter Rousseeuw, KU Leuven, Belgium and Prof. Patrick J. Wolfe, Purdue University, United States.

The organization would like to thank the authors, referees and all participants of COMPSTAT 2022 who contributed to the success of the conference. Our gratitude to sponsors, the scientific programme committee, session organizers, local hosts, the city of Bologna, and many volunteers who have contributed substantially to the conference. We acknowledge their work and support.

The forthcoming COMPSTAT conference, which has been affected by the postponement, will take place in an odd year as an exception. The COMPSTAT 2023 organizers invite you to participate in London, UK, 22-25 August 2023. We wish the best of success to Erricos Kontoghiorghes, Chair of the 25th COMPSTAT edition.

Alessandra Luati and Maria Brigida Ferraro.

SCHEDULE

COMPSTAT 2022

2022-08-23	2022-08-24	2022-08-25	2022-08-26
A - Keynote 09:00 - 10:00	F 09:00 - 10:30	I 09:00 - 11:00	M 09:00 - 10:30
Coffee Break 10:00 - 10:30	Coffee Break 10:30 - 11:00	Coffee Break 11:00 - 11:30	Coffee Break 10:30 - 11:00
B 10:30 - 12:30	G 11:00 - 12:30	J - Keynote 11:30 - 12:20	N 11:00 - 12:00
Lunch Break 12:30 - 14:15	Lunch Break 12:30 - 14:15	Lunch Break 12:20 - 14:15	O - Keynote 12:10 - 13:15
C 14:15 - 15:45	H 14:15 - 16:15	K 14:15 - 15:45	
Coffee Break 15:45 - 16:15		Coffee Break 15:45 - 16:15	
D 16:15 - 17:45		L 16:15 - 17:45	
E 17:55 - 18:55		Conference Dinner 19:30 - 22:30	
Welcome reception 19:10 - 20:40			

SCHEDULE

SDS 2022

2022-08-26

- **A - Keynote** 15:00 - 16:00
- **Coffee Break** 16:00 - 16:30
- **B** 16:30 - 18:10
- **Welcome reception** 18:15 - 19:45

2022-08-27

- **C** 09:00 - 10:15
- **Coffee Break** 10:15 - 10:45
- **D** 10:45 - 12:25
- **Lunch Break** 12:25 - 13:55
- **E - Keynote** 13:55 - 14:45
- **F** 14:55 - 16:35
- **Coffee Break** 16:35 - 17:05
- **G** 17:05 - 18:45
- **Conference Dinner** 20:00 - 22:30

2022-08-28

- **H** 09:00 - 11:05
- **Coffee Break** 11:05 - 11:35
- **I - Keynote** 11:35 - 12:30

TUTORIALS, MEETINGS AND SOCIAL EVENTS

TUTORIALS - COMPSTAT 2022

The tutorials will take place in room Aula G, Plesso Belmeloro, in parallel with the invited, organized and contributed sessions. The first one is given by Prof. Peter Winker (*Introductionary tutorial to text mining in econometrics*), Tuesday 23.8.2022, 10:30 - 12:30. The second is given by Prof. Fabrizio Durante (*Tail dependence with copulas*), Thursday 25.8.2022, 09:00 - 11:00.

SPECIAL MEETINGS by invitation to group members

- IASC Executive Committee meeting, *Room: Aula G, Plesso Belmeloro*, Tuesday 23rd August 2022, 12:35-14:05.
- ERS BoD Meeting, *Room: Aula G, Plesso Belmeloro*, Wednesday 24th August 2022, 12:35-14:05.
- IASC and ERS General Assembly, *Room: Aula G, Plesso Belmeloro*, Thursday 25th August 2022, 17:50-18:50.

SOCIAL EVENTS - COMPSTAT 2022

- *The coffee breaks* will take place in the ground-level Hall of Plesso Belmeloro, building A. You must have your conference badge in order to attend the coffee breaks.
- *Sandwich Lunch Boxes* will be available 23-26 August 2022 for those who had booked. The lunches are optional and registration is required. Information about the purchased lunch tickets is embedded in the QR code on the conference badge. You must have your conference badge in order to get your lunch each day in the ground-level Hall of Plesso Belmeloro, building A. People not registered for lunch can buy lunch at restaurants and cafes in close walking distance to the conference venue.
- *Welcome Reception, Tuesday 23rd August 2022t, 19:10-20:40.* The Welcome Reception will take place at the Palazzo Poggi, and is open to all registrants who had preregistered and accompanying persons who have purchased a reception ticket. Participants must bring their conference badges in order to attend the reception. Preregistration is required due to health and safety reasons.
- *Food Walking Tour, Wednesday 24th August 2022, 16:30-19:00.* Walking tour through the streets of Bolognas market, tasting local specialities and discovering curiosities about Bolognese food. This tour will allow everyone to explore the gastronomy of Bologna with a walk inside the Quadrilatero market, the real beating heart of local products. Duration: 2 hours 30 minutes. The visit is optional, and registration is required. Participants must bring their conference badges in order to attend the event. Information about the booking is embedded in the QR code on the conference badge. The meeting point is the the entrance of the Palazzo Poggi at Via Zamboni.
- *Conference Dinner, Thursday 25th August 2022, 19:30-22:30.* The conference dinner will take place at the Royal Hotel Carlton. It is optional and registration is required. Before the dinner, there will be a reception in the garden. Participants must bring their conference badges in order to attend the conference dinner. Information about the dinner booking is embedded in the QR code on the conference badge.

SOCIAL EVENTS - SDS 2022

- *The coffee breaks* will take place in room Aula 2, ground floor of the Department of Statistical Sciences "Paolo Fortunati". You must have your SDS 2022 badge in order to attend the coffee breaks.
- *Welcome Reception, Friday 26th August 2022,, 18:15-19:45.* The Welcome Reception will take place in the inner garden of the Department of Statistical Sciences "Paolo Fortunati". It is free for all registrants. Registrants must bring their badges in order to attend the reception.
- *Satellite Workshop Dinner, Saturday 27th August 2022, 20:00-22:30.* The SDS 2022 dinner is optional and registration is required. It will take place at 'the Royal Hotel Carlton. SDS 2022 registrants must bring their conference badges in order to attend the dinner.

GENERAL INFORMATION (see maps on pages IX to XII)

Address of venues

The Conference venue is the Plesso Belmeloro of the University of Bologna, via Beniamino Andreatta 14, Bologna, Italy. The satellite events SDS venue is the Department of Statistical Sciences "Paolo Fortunati", University of Bologna, via delle Belle Arti, n.41, 40126 Bologna, Italy Alexandru I.

Registration

The registration will be open during the conference days from 08:40 to 17:00 and will take place in the ground-level Hall of Plesso Belmeloro, building A. On Friday 26th August at 13:00, it will be moved to room Aula 2 on the ground floor of the Department of Statistical Sciences "Paolo Fortunati".

Presentation instructions

The opening, keynote and closing talks will take place in room Aula B. The COMPSTAT parallel sessions will take place in rooms Aula B, C, D, E, F, G, H, I and Q. In-person participants can use room Aula P as a quiet room to attend virtual sessions from their own devices. The poster sessions will take place online, but in-person participants are invited to meet in Aula P with their own laptops for related discussions. The virtual presentations will take place through Zoom. Speakers should have a stable internet connection, and ensure their video and audio are working. They will share their slides when the chair requires it, present their talk, and answer questions after the presentation. The in-person speakers must share presentations through the Zoom session open on the desktop in the conference rooms. The rooms have a webcam, and an omnidirectional desk microphone that collects the sound around the PC desk to make the live streaming easy. Detailed indications for speakers in either virtual or hybrid sessions can be found on the website. As a general rule, each speaker has 20 minutes, including 2-3 minutes for discussion. Strict timing must be observed.

Posters

The poster sessions will take place through Gather Town. The posters should be sent in png format to info@compstat2022s.org by the 19th of August. Landscape orientation is advisable. Detailed indications for the poster presentations can be found on the website.

Session chairs

The session chairs will be responsible for introducing the session, the speakers and coordinating the discussion time. A member of the conference staff, identified on Zoom by the name Angel followed by a number, will assist online. If any speaker is missing or has a technical problem, the chair can pass to the next speaker and come back later to resume if possible. Detailed indications for the session chairs of both virtual and hybrid sessions can be found on the website.

Test session

A test session will be set up for Saturday 20th August 2022, from 15:00 to 15:30 GMT+2. The participants will be able to enter the virtual room Aula B in the programme to test their presentations, video, micro and audio. Detailed indications for the test sessions can be found on the website.

Map of the COMPSTAT venue

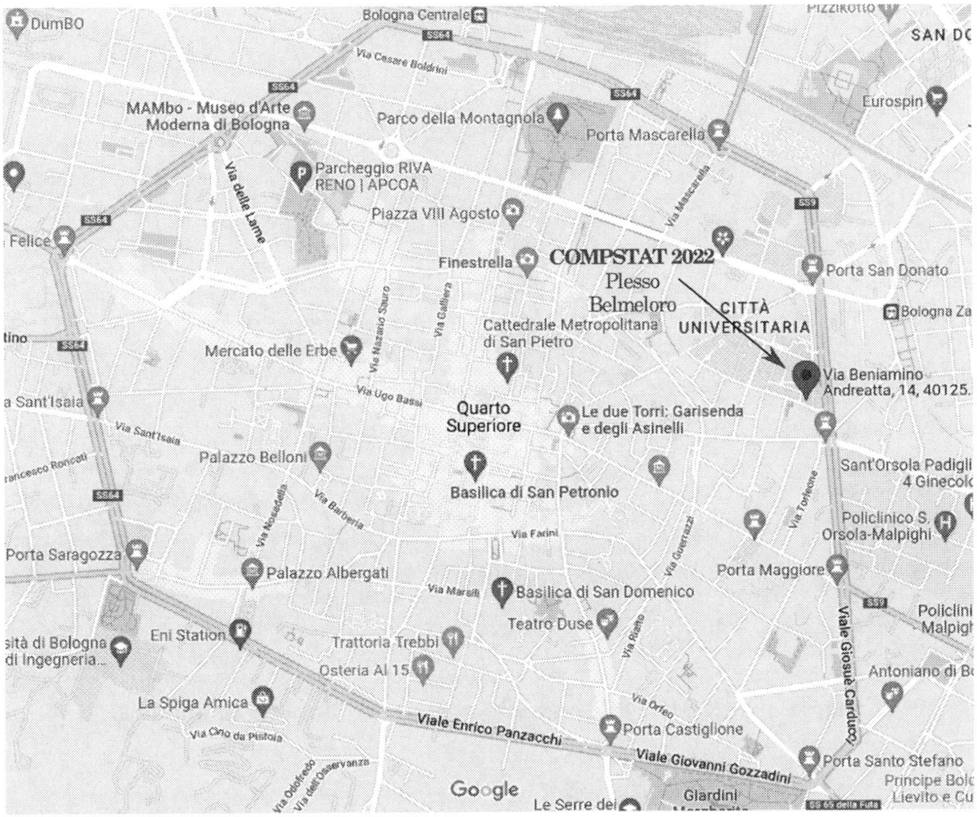

Map of the SDS venue and SDS welcome reception

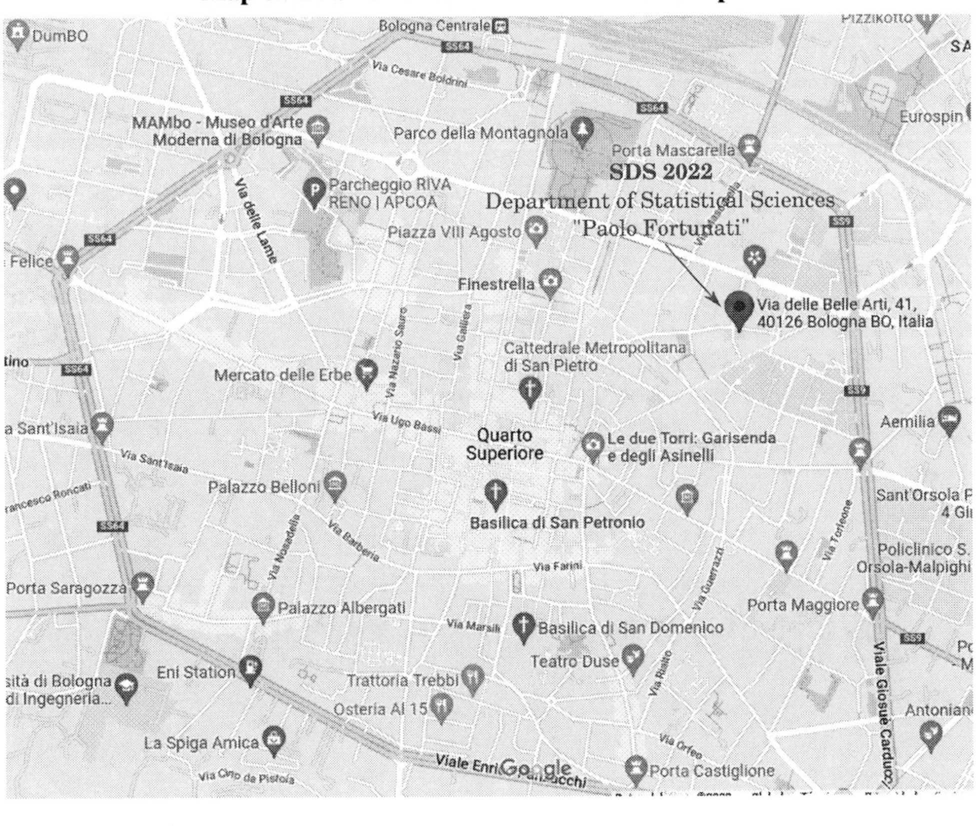

Map of the COMPSTAT welcome reception (5 minutes)

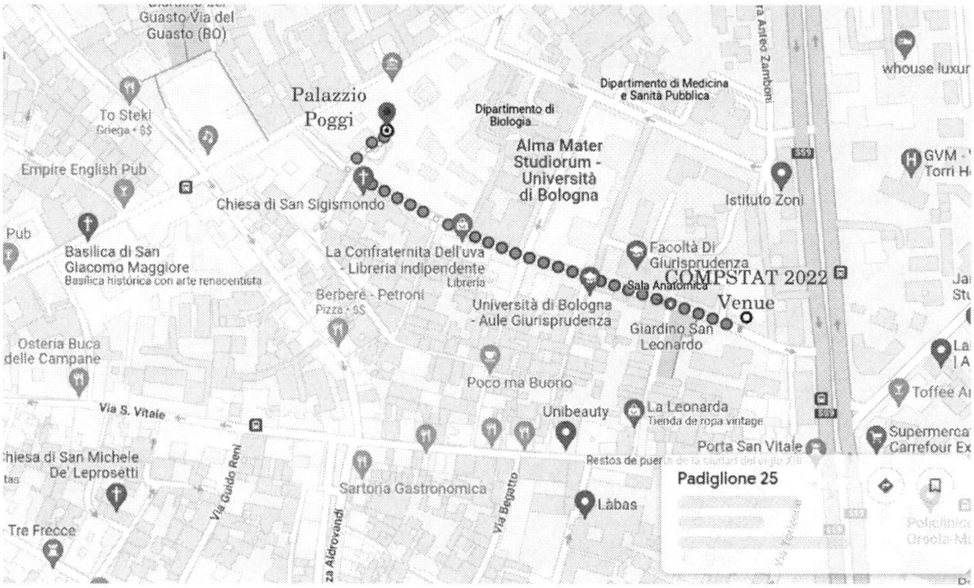

Map of the COMPSTAT and SDS conference dinners (20 minutes)

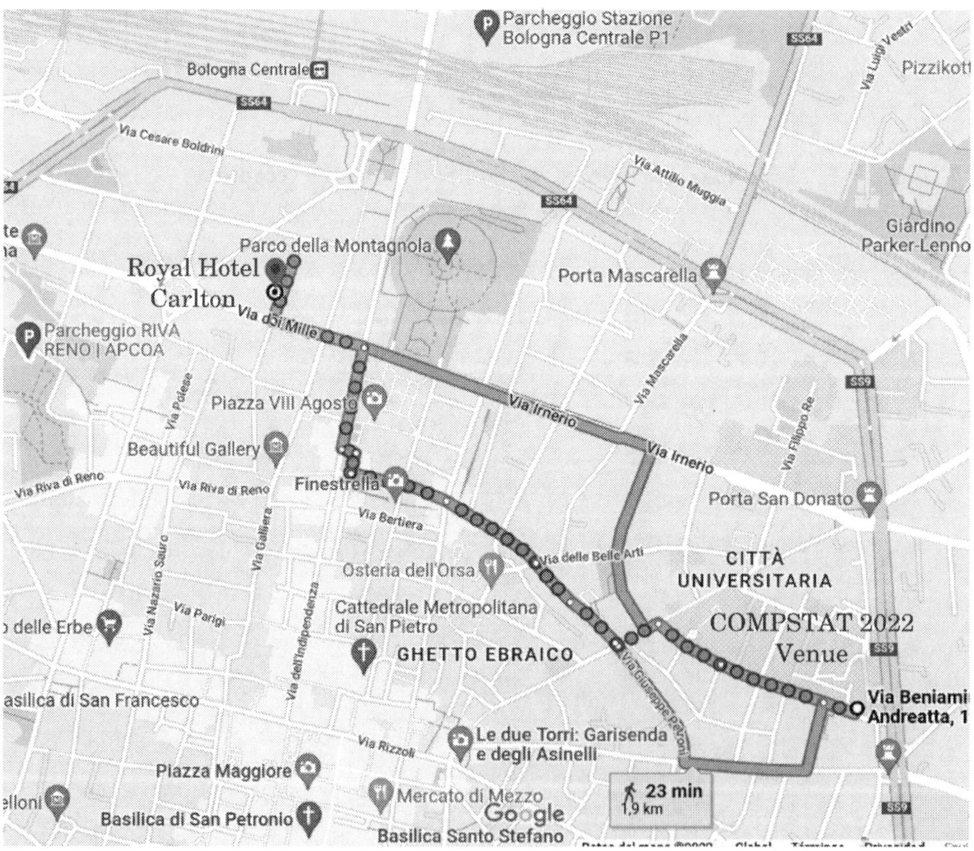

COMPSTAT venue - Plesso Belmeloro, Ground Floor

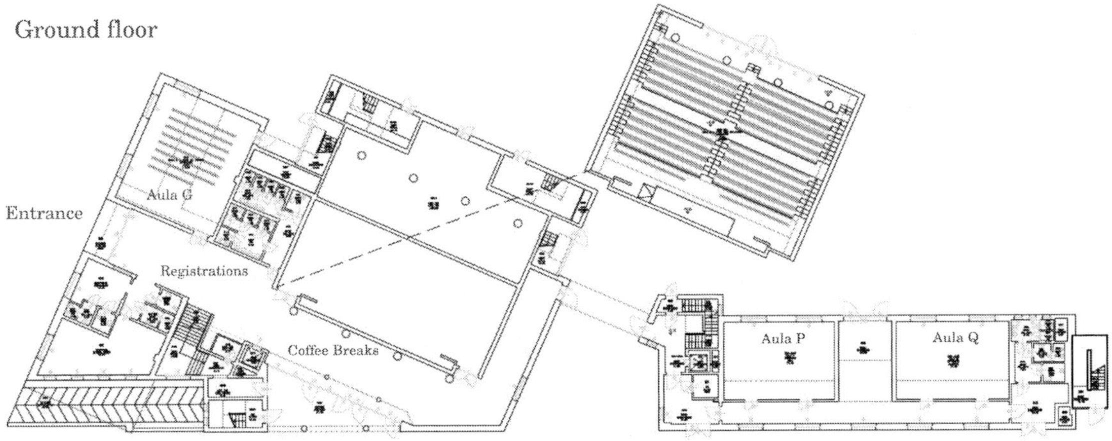

COMPSTAT venue - Plesso Belmeloro, First Floor

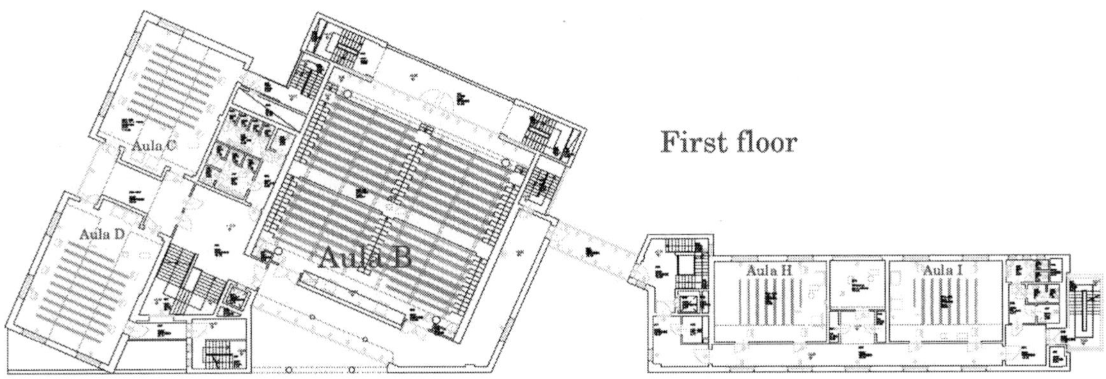

COMPSTAT venue - Plesso Belmeloro, Second Floor

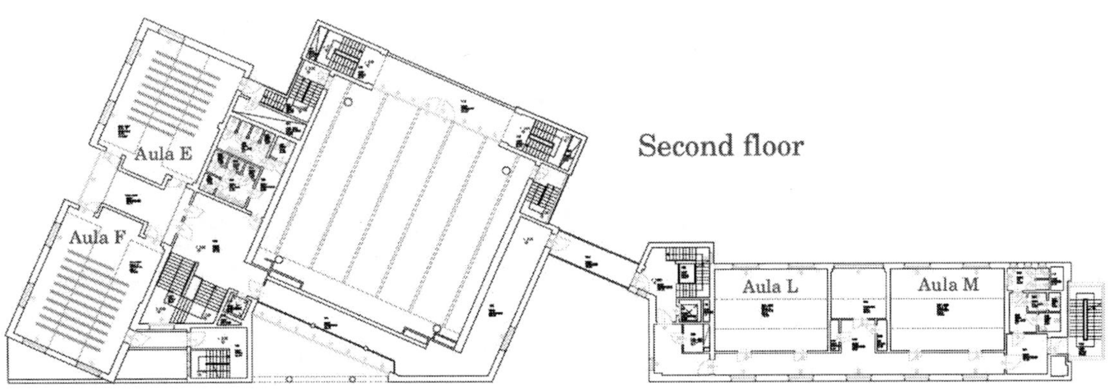

SDS venue - Department of Statistical Sciences "Paolo Fortunati", Ground Floor

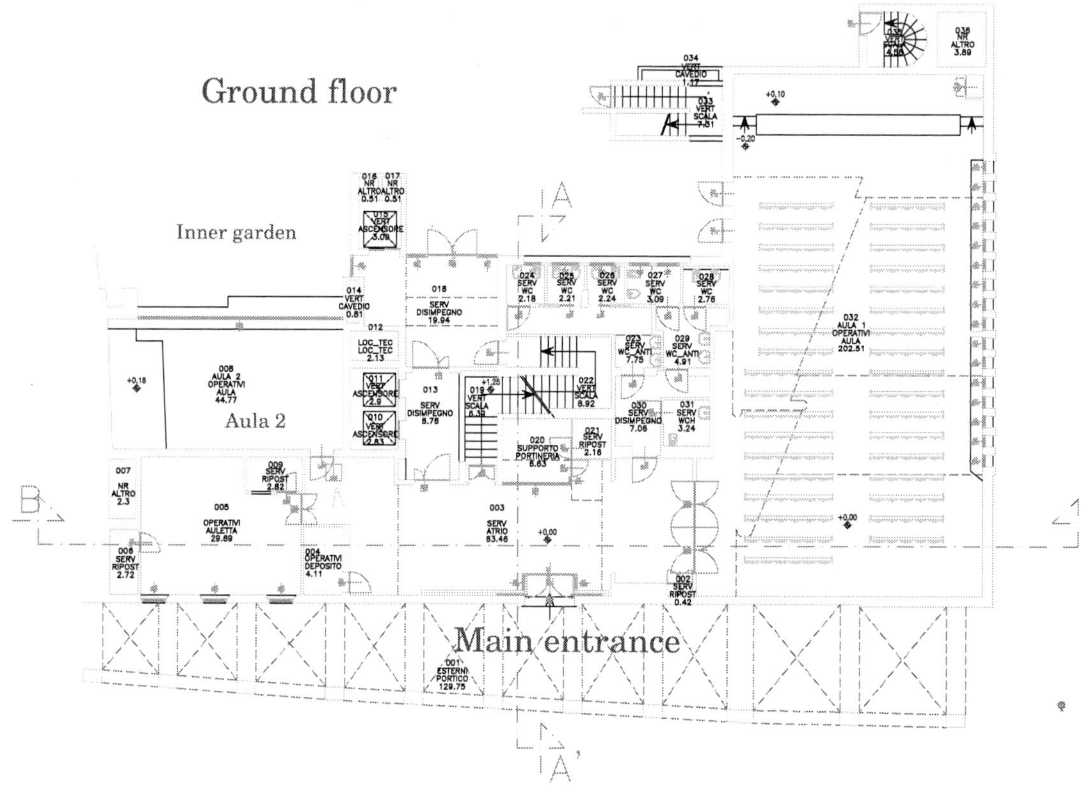

SDS venue - Department of Statistical Sciences "Paolo Fortunati", Second Floor

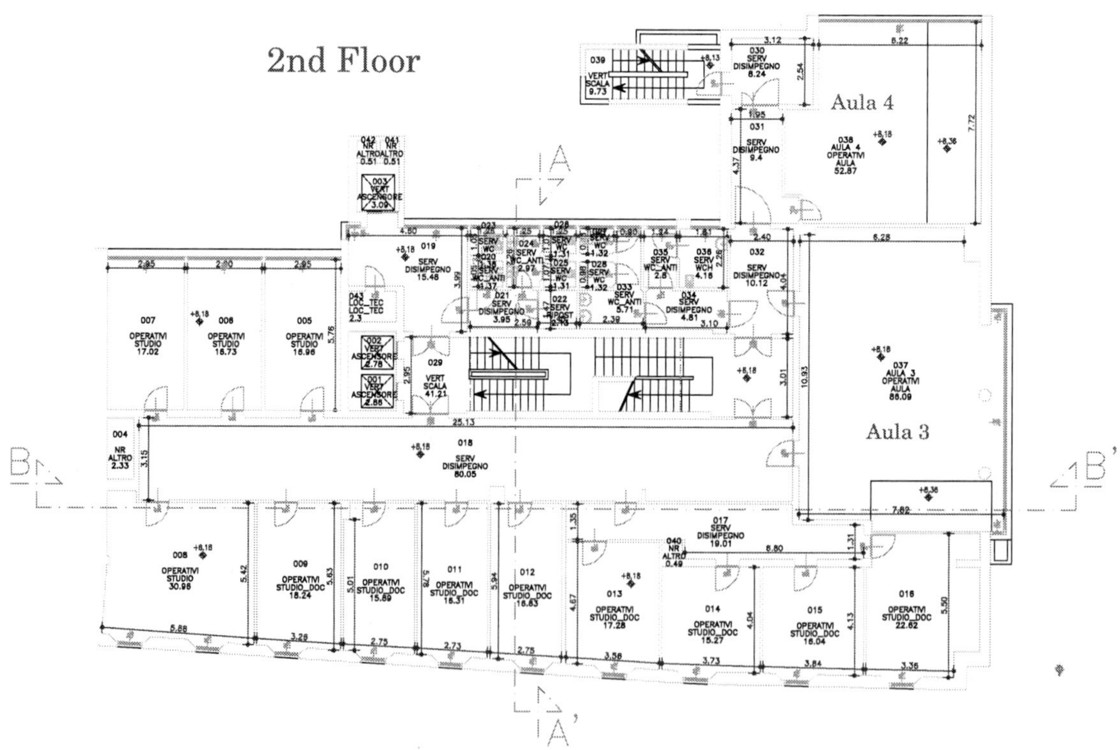

Contents

Tuesday 23.08.2022	09:00 - 10:00	Room: Aula B	Chair: Maria Brigida Ferraro	Keynote talk 1

Are deviations in a gradually varying mean relevant?

Speaker: **Holger Dette, Ruhr-Universitaet Bochum, Germany**

Classical change point analysis aims at (1) detecting abrupt changes in the mean of a possibly non-stationary time series and at (2) identifying regions where the mean exhibits a piecewise constant behaviour. In many applications (for example climatology) however, it is more reasonable to assume that the mean changes gradually in a smooth way. Those gradual changes may either be non-relevant (i.e., small), or relevant for a specific problem at hand. We develop a statistical methodology to detect the latter. More precisely, we consider a locally stationary process with a time-varying trend and propose a test for the null hypothesis that the maximum absolute deviation of the trend from a given benchmark (such as the value of the trend at the beginning of the observation period or an average value over the past) is smaller than a given threshold. A test for this type of hypothesis is developed using an appropriate estimator for the maximum deviation. We also provide estimates for the time point, where this relevant deviation appears for the first time.

Thursday 25.08.2022	11:30 - 12:20	Room: Aula B	Chair: Alessandra Luati	Keynote talk 2

Testing the existence of moments and estimating the tail index of augmented GARCH processes

Speaker: **Jean-Michel Zakoian, CREST, France** Christian Francq

The aim is to investigate the problem of testing the finiteness of moments for a class of semi-parametric augmented GARCH models encompassing the most commonly used specifications. The existence of positive-power moments of the strictly stationary solution is characterized through the Moment Generating Function (MGF) of the model, defined as the MGF of the logarithm of the random autoregressive coefficient in the volatility dynamics. We establish the asymptotic distribution of the empirical MGF, from which tests of moments are deduced. Alternative tests relying on the estimation of the Maximal Moment Exponent (MME) are studied. Power comparisons based on local alternatives and the Bahadur approach are proposed. We provide illustrations of Monte Carlo experiments and real financial data, showing that semi-parametric estimation of the MME offers an interesting alternative to Hill's nonparametric estimator of the tail index.

Friday 26.08.2022	12:10 - 13:15	Room: Aula B	Chair: Christophe Croux	Keynote talk 3

Learning and prediction via hierarchies of random measures in Bayesian nonparametrics

Speaker: **Igor Pruenster, Bocconi University, Italy**

The Hierarchical Dirichlet Process enjoyed huge success in the MachineLearning literature. It originated in the context of topic modeling as a powerful generalization of the ubiquitous Latent Dirichlet Allocation model that allows learning the number of topics from the data. the hierarchical Dirichlet Process and more general Bayesian nonparametric models constructed as hierarchies of random probability measures can be naturally embedded within the framework of partial exchangeability, a neat probabilistic representation for multiple distinct yet related populations. Moreover, the discrete nature of these models yields ties across populations, resulting in a shrinkage property often described as sharing of information" and crucial for the learning mechanism. We obtain distributional results, which include a characterization of the induced random partitions and a complete posterior representation: these are key to the derivation of effective sampling schemes. Further interesting extensions deal with tree structures and hierarchies of random measures. Illustrations concerning species sampling, survival analysis, network theory and topic modeling are provided.

Tuesday 23.08.2022 10:30 - 12:30 Parallel Session B – COMPSTAT2022

CI013 Room Aula B SMALL AREA ESTIMATION **Chair: Stefan Sperlich**

C0604: **Empirical best prediction of bivariate nonlinear small area indicators**
Presenter: **Domingo Morales**, University Miguel Hernandez of Elche, Spain
Co-authors: Maria-Dolores Esteban, Maria Jose Lombardia, Esther Lopez Vizcaino, Agustin Perez

The small area estimation of population averages of unit-level compositional data is investigated. The new methodology transforms the compositions into vectors of a Euclidean space and assumes that the vectors follow a multivariate nested error regression model. Empirical best predictors of domain indicators are derived from the fitted model and their mean squared errors are estimated by parametric bootstrap. The empirical analysis of the behaviour of the introduced predictors is investigated by means of simulation experiments. An application to real data from the Spanish household budget survey, in 2016, is given. The target is to estimate the average proportions of annual household expenditures on food, housing and others, by Spanish provinces.

C0309: **Full bias correction approaches for M-quantile small area estimators: Application to Italian labour force survey**
Presenter: **Francesco Schirripa Spagnolo**, Universita di Pisa - Dipartimento di Economia e Management, Italy
Co-authors: Gaia Bertarelli, Raymond Chambers, David Haziza, Nicola Salvati

When representative outlier units are a concern for the estimation of population quantities, it is essential to pay attention to them in a small area estimation (SAE) context. Standard approaches use plug-in robust prediction replacing parameter estimates in optimal but outlier-sensitive predictors with outlier robust versions. These predictors are efficient under the correct model but may be sensitive to the presence of outliers because they use plug-in robust prediction which usually leads to a low prediction variance and a considerable prediction bias. We propose two new full bias correction methods to reduce the prediction bias of the robust M-quantile predictors in SAE for continuous, count and binary data. The first estimator is based on the concept of conditional bias. The second one is based on a full bias correction. The properties of the proposed estimators are empirically assessed in model-based and design-based simulations. These estimators correct for the bias and are more efficient than the robust-predictive estimators and robust-projective estimators in the presence of area and individual outliers. Two estimators of the prediction mean-squared error are described. The methodology proposed is applied to Italian annual Labour Force Survey data for estimating the proportion of the unemployed in local labour market areas.

C0428: **Variable selection using conditional AIC for linear mixed models with data-driven transformations**
Presenter: **Yeonjoo Lee**, Otto-Friedrich-Universitat Bamberg, Germany
Co-authors: Marina Runge, Natalia Rojas Perilla, Timo Schmid

When data analysts use linear mixed models, they usually encounter two practical problems: a) the true model is unknown and b) the Gaussian assumptions of the errors do not hold. While these problems commonly appear together, researchers tend to treat them individually by a) finding an optimal model based on the conditional Akaike information criterion (cAIC) and b) applying transformations on the dependent variable. However, the optimal model depends on the transformation and vice versa. We aim to solve both problems simultaneously. In particular, we propose an adjusted cAIC by using the Jacobian of the particular transformation such that various model candidates with differently transformed data can be compared. From a computational perspective, we propose a step-wise selection approach based on the introduced adjusted cAIC to reduce computational costs. Model-based simulations are used to compare the proposed selection approach to alternative approaches. Finally, the introduced approach is applied to Mexican data to estimate poverty and inequality indicators for 81 municipalities.

C0180: **Uniform inference for SAE**
Presenter: **Stefan Sperlich**, University of Geneva, Switzerland
Co-authors: Maria Jose Lombardia, Katarzyna Reluga

Simultaneous inference is addressed for mixed parameters which are the key ingredients in small area estimation. We assume linear mixed model framework. We analyse statistical properties of a max-type statistic and use it to construct simultaneous prediction intervals as well as to implement multiple testing procedure. In addition, we adapt some of the simultaneous inference methods from regression and nonparametric curve estimation and compare them with our approaches. Simultaneous intervals are necessary to compare areas since the presently available intervals are not statistically valid for such analysis. The proposed testing procedures can be used to validate certain statements about the set of mixed parameters or to test pairwise differences. The proposal is accompanied by simulation experiments and a data example on small area household incomes. They all demonstrate an excellent performance and utility of our techniques.

CO166 Room Aula G TUTORIAL I **Chair: Peter Winker**

C0472: **Introductory tutorial to text mining in econometrics**
Presenter: **Peter Winker**, University of Giessen, Germany

There is a growing interest in and use of textual information in different fields of economics comprising financial markets (statements by analysts, communication of central banks) over innovation activities (patent abstracts, firm websites), and the development of economic science (journal articles). Using such textual information for quantitative analysis involves several steps including, e.g., 1) the selection of appropriate sources (corpora) and establishing access, 2) the preparation of the text data for further analysis, 3) the identification of themes within documents, 4) the quantification of the relevance of themes in different documents, 5) the aggregation of relevant information, e.g., across sectors or over time, 6) the application of the generated indicators. The tutorial will provide some first insights and recommendations concerning these steps of the analysis and address open issues regarding, e.g., computational complexity and statistical robustness of the methods. All steps will be illustrated with empirical examples.

CO073 Room Aula D STATISTICAL ANALYSIS IN FINITE AND INFINITE DIMENSIONAL HILBERT SPACES **Chair: Karel Hron**

C0328: **Weighting of parts in compositional data using Bayes Hilbert spaces**
Presenter: **Karel Hron**, Palacky University, Czech Republic
Co-authors: Alessandra Menafoglio, Javier Palarea-Albaladejo, Peter Filzmoser, Juan Jose Egozcue

It often occurs in practice that it is sensible to give different weights to the variables involved in multivariate data analysis. The same holds for compositional data as multivariate observations carrying relative information, such as proportions or percentages. It can be convenient to apply weights to, for example, better accommodate differences in the quality of the measurements, the occurrence of zeros and missing values, or generally to highlight some specific features of compositional variables (i.e. parts of a whole). The characterisation of compositional data as elements of a Bayes space with the Hilbert space structure enables the definition of a formal framework to implement weighting schemes for the parts of a composition. This is formally achieved by considering a reference measure in the Bayes space alternative to the common uniform measure via the well-known chain rule. Unweighted centred logratio (clr) coefficients and isometric logratio (ilr) coordinates then allow us to represent compositions in the real space equipped with the (unweighted) Euclidean geometry, where ordinary multivariate statistical methods can be used and interpreted. We present these formal developments and use them to introduce a general approach to weighting parts in compositional data analysis. We demonstrate its practical usefulness on simulated and real-world data sets in the context of the earth sciences.

C0356: **Analysis of multi-factorial compositional data main principles and perspectives**
Presenter: **Kamila Facevicova**, Palacky University Olomouc, Czech Republic
Co-authors: Peter Filzmoser, Karel Hron

Compositional data are in their traditional setting understood as vector observations of positive entries. This means, that the observations carry information about the relative structure given by levels of one factor. The contribution will focus on a more complex situation where the structure is given by two or more determining factors. The two-factorial case is already thoroughly described in the literature and can be found under the keyword compositional tables. It turns out that the main findings of the geometrical structure of tables and their coordinate representation can be further extended to the multi-factorial case. The presentation brings an overview of the current state of knowledge in the field of analysis of multi-factorial compositional data. The theoretical principles will be followed by practical examples and, finally, perspectives of the further research in the field will be discussed.

C0319: **Additive density-on-scalar regression in Bayes Hilbert spaces with an application to gender economics**
Presenter: **Sonja Greven**, Humboldt University of Berlin, Germany
Co-authors: Eva-Maria Maier, Almond Stoecker, Bernd Fitzenberger

Motivated by research on gender identity norms and the distribution of the woman's share in a couple's total labor income, functional additive regression models for probability density functions as responses with scalar covariates are considered. To preserve nonnegativity and integration to one under summation and scalar multiplication, we formulate the model for densities in a Bayes Hilbert space with respect to an arbitrary finite measure. This enables us to not only consider continuous densities but also, e.g., discrete or mixed densities. Mixed densities occur in our application, as the woman's income share is a continuous variable having discrete point masses at zero and one for single-earner couples. We discuss the interpretation of effect functions in our model via odds-ratios. Estimation is based on a gradient boosting algorithm, allowing for potentially numerous flexible covariate effects. We show how to handle the challenging estimation for mixed densities within our framework using an orthogonal decomposition. Applying this approach to data from the German Socio-Economic Panel Study (SOEP) shows a more symmetric distribution in East German than in West German couples after reunification and a smaller child penalty comparing couples with and without minor children. These West-East differences become smaller but are persistent over time.

C0352: **Scalar-on-function regression and functional isotemporal substitution analysis in the context of time-use data**
Presenter: **Paulina Jaskova**, Palacky University Olomouc, Czech Republic
Co-authors: Karel Hron, Javier Palarea-Albaladejo, Ales Gaba, Zeljko Pedisic, Dorothea Dumuid

How people allocate their daily time to physical activities (PA), sedentary behaviors (SB) and sleep have important health implications. PA (categorized by intensity into light, moderate and vigorous PA), SB and sleep should not be analyzed separately, because they are parts of a time-use composition with a natural constraint of 24 hours per day. To find out how relative reallocations of time between PA of various intensities are associated with health, we describe compositional scalar-on-function regression (CSFR) and a newly developed functional isotemporal substitution analysis (FISA). Instead of working with the ordinary PA categorization, the original data consisting of PA intensity distributions can be characterized as empirical probability density functions (PDFs), which better reflect the continuous character of their measurement using accelerometers. These PDFs have specific properties, such as scale invariance and relative scale, and they are geometrically represented using Bayes spaces. The CSFR of explanatory PDFs and FISA were applied to a dataset from a cross-sectional study on 24-h movement behaviors and adiposity conducted among school-aged children in the Czech Republic. Theoretical reallocations of time to PA of higher intensities were found to be associated with larger decreases in adiposity. We gained detailed insight into the dose-response relationship between PA intensity and adiposity, which would not be feasible using an ordinary approach.

C0414: **Data driven orthogonal basis selection for functional data analysis**
Presenter: **Hiba Nassar**, Technical University of Denmark, Denmark
Co-authors: Krzysztof Podgorski, Rani Basna

Functional data analysis is typically performed in two steps: first, functionally representing discrete observations, and then applying functional methods, such as the functional principal component analysis, to the so-represented data. While the initial choice of a functional representation may have a significant impact on the second phase of the analysis, this issue has not gained much attention in the past. Typically, a rather ad hoc choice of some standard basis such as Fourier, wavelets, splines, etc. is used for the data transforming purpose. To address this important problem, we present its mathematical formulation, demonstrate its importance, and propose a data-driven method of functionally representing observations. The method chooses an initial functional basis by an efficient placement of the knots. A simple machine learning style algorithm is utilized for the knot selection and recently introduced orthogonal spline bases - splinets - are eventually taken to represent the data. The benefits are illustrated by examples of analyses of sparse functional data.

CO142 Room Aula Q ALGEBRAIC STATISTICS	**Chair: Marta Nai Ruscone**

C0372: **Geometry of memoryless policy optimization in POMDPs**
Presenter: **Guido Montufar**, UCLA and MPI MiS, Germany
Co-authors: Johannes Mueller

The focus is on the problem of finding the best memoryless stochastic policy for an infinite-horizon partially observable Markov decision process (POMDP) with finite state and action spaces with respect to either the discounted or mean reward criterion. We show that the (discounted) state-action frequencies and the expected cumulative reward are rational functions of the policy, whereby the degree is determined by the degree of partial observability. We then describe the optimization problem as a linear optimization problem in the space of feasible state-action frequencies subject to polynomial constraints that we characterize explicitly. This allows us to address the combinatorial and geometric complexity of the optimization problem using tools from polynomial optimization. In particular, we estimate the number of critical points and use the polynomial programming description of reward maximization to solve a navigation problem in a grid world.

C0236: **Recent developments in hybrid causal discovery**
Presenter: **Liam Solus**, KTH Royal Institute of Technology, Sweden

Combinatorics, algebra and discrete geometry have come to play an increasingly significant role in the development of methods for causal discovery, the goal of which is to infer the cause-effect relations amongst a system of variables based on available data. Since the basic model selection problem of causal discovery is NP-hard, a variety of techniques for learning a causal model efficiently have been studied. One family of such methods are the hybrid causal discovery algorithms, which rely on a mixture of conditional independence testing and greedy optimization. We will discuss some recently explored connections between causal discovery and classic problems in combinatorial optimization that yield hybrid algorithms with improved performance over the state-of-the-art. We will see that the geometric perspective promoted by fields such as algebraic statistics plays a key role in the identification of this new methodology.

C0663: **Identifiability in continuous graphical Lyapunov models**
Presenter: **Carlos Amendola**, Technical University Berlin, Germany
Co-authors: Philipp Dettling, Mathias Drton, Niels Richard Hansen, Roser Homs

Lyapunov graphical models represent a new approach in graphical modeling where independent observations are taken to be one-time cross-sectional snapshots of the multivariate Ornstein-Uhlenbeck process in equilibrium. The non-zero pattern of the drift matrix allows for a causally interpretable dependence structure among the coordinates of the process which can be represented by a directed graph. We will introduce these models and focus on the fundamental question of identifiability, i.e. being able to recover the parameters knowing the true data generating distribution. Familiar algebraic tools such as matrix rank computations and the sum of squares decompositions help us study this identifiability problem for both directed acyclic and simple cyclic graphs.

C0497: **Multivariate Bernoulli distributions and discrete copulas**

Presenter: **Elisa Perrone**, Eindhoven University of Technology, Netherlands

Co-authors: Roberto Fontana

Multivariate Bernoulli distributions are used in many applied domains such as healthcare, social sciences, and finance. The class of d-dimensional Bernoulli distributions, with given Bernoulli univariate marginal distributions, admits a representation as a convex polytope. For exchangeable multivariate Bernoulli distributions with given margins, an analytical expression of the extreme points of the polytope has recently been determined. Discrete copulas are statistical tools to represent the joint distribution of discrete random vectors. They are fascinating mathematical objects that also admit a representation as a convex polytope. Studying polytopes of discrete copulas and their extreme points has recently gained attention in the literature. We explore potential connections between multivariate Bernoulli distributions, discrete copulas, and the extreme points of their associated polytopes. We discuss possible ways to unify the literature on the topics and describe some numerical examples.

C0332: **Markov bases from discrete to continuous frameworks**

Presenter: **Fabio Rapallo**, University of Genova, Italy

The notions of Markov moves and Markov bases from Algebraic Statistics are traditionally defined in a discrete framework to define a connected Markov chain on the set of contingency tables under linear constraints, thus working with empirical distributions of counts. In the context of rater agreement analysis, Markov bases can be used to find the maximum value of kappa-type statistics through a simulated annealing algorithm. The same objects and the same algorithm can also be applied in a continuous setting, i.e., working with probability distributions, to find the Kantorovich distance between two distributions on a finite sample space. Here we highlight the analogies and the differences between the two approaches.

CC158 **Room Aula C** **TIME SERIES** **Chair: Rob Hyndman**

C0297: **Random forests for time series**

Presenter: **Jean-Michel Poggi**, University Paris-Saclay Orsay, France

Co-authors: Yannig Goude, Hui Yan, Benjamin Goehry, Pascal Massart

Random forests were introduced in 2001 by Breiman and have since become a popular learning algorithm, for both regression and classification. However, when dealing with time series, random forests do not integrate the time-dependent structure, implicitly supposing that the observations are independent, and treat each instant as an independent observation. We propose some variants of the random forests designed for time series. The idea is to replace the standard bootstrap with a dependent bootstrap (i.e. block bootstrap) to subsample time series during the tree construction phase and take time dependence into account. We then present two numerical experiments on electricity load forecasting. The first one, at a disaggregated level, is based on an application to load forecasting of a building and illustrate how the variants may perform. The second one is at a more aggregated level (French national forecasting) but focusing on atypical periods. For both, we explore a heuristic for the choice of the block size, the new parameter. In addition, some additional experiments with generic time series data are also performed and shortly commented. Finally, our R package rangerts is freely available from the GitHub.

C0299: **Decomposing time series with complex seasonality**

Presenter: **Rob Hyndman**, Monash University, Australia

Time series data often contain a rich complexity of seasonal patterns. Time series that are observed at a sub-daily level can exhibit multiple seasonal patterns corresponding to different granularities such as hour-of-the-day, day-of-the-week or month-of-the-year. They can be nested (e.g., hour-of-the-day within day-of-the-week) and non-nested (e.g., day-of-the-year in both the Gregorian and Hijri calendars). We will discuss two new time series decomposition tools for handling seasonalities in time series data: MSTL and STR. These allow for multiple seasonal and cyclic components, covariates, seasonal patterns that may have non-integer periods, and seasonality with complex topology. They can be used for time series with any regular time index including hourly, daily, weekly, monthly or quarterly data, but tackle many more decomposition problems than other methods allow.

C0390: **Testing for and measuring serial dependence by neural networks**

Presenter: **Jinu Lee**, King's College London, United Kingdom

Testing serial dependence is central to much of time series econometrics. The focus is on the generalisation of an autocorrelation function to test for and measure serially dependent processes by using neural networks based approximations. Simulations find that the suggested nonparametric method shows good power properties and has the potential to measure nonlinear associations compared to some popular tests and measures. An application to US stock returns illustrates the usefulness of the proposed tests and measures for nonlinear dependences.

C0420: **Multivariate adaptive learning forecasting**

Presenter: **Foteini Kyriazi**, Agricultural University of Athens, Greece

Co-authors: Dimitrios Thomakos, John Guerard

A new method is presented for forecasting multivariate time series, either in simultaneous equations or panel form, that utilizes adaptive learning on past forecast errors for effecting root mean-squared error reductions to any input forecast that one wishes to utilize. The method is the multivariate extension of univariate adaptive learning and presents a number of significant advantages over the univariate approach, as most multivariate methods do when dealing with related time series. The multivariate version of adaptive learning can be used in a number of different settings and can be used to extract useful information for forecasting from different forms of covariation among multiple time series. We explore in detail the theoretical foundations of the method, and the computational requirements and present a number of simulation and empirical examples that illustrate both the efficacy and the competitiveness of the method compared to a number of well-known time series benchmarks.

C0528: **A changepoint approach to modelling soil moisture dynamics**

Presenter: **Mengyi Gong**, Lancaster University, United Kingdom

Co-authors: Rebecca Killick, Christopher Nemeth

Soil moisture is an important measure of soil health that scientists model via soil drydown curves. The typical modelling process requires manually identifying the drying process and fitting exponential decay models to them. This can be time-consuming and the result is a static overview of the drydown property. Motivated by the spike-train problem in neuroscience, a novel changepoint-based approach is proposed to automatically identify structural changes in the soil drying process. Changes caused by sudden rises in soil moisture content over a long time series are captured and the parameters characterising the drying processes are estimated simultaneously. Segment-specific parameters are used to capture potential temporal variations in the drying process. An algorithm based on the penalised exact linear time (PELT) method was developed to identify the changepoints. Applying the algorithm to simulated and real data show the good performance of the method. To improve flexibility, the method is extended such that different types of models can be used to describe different segments. This allows the segmentation of the time periods when no drydown happens due to low temperature or saturation, in addition to the typical drying periods. An approach based on the Bayesian changepoint detection method and the particle Metropolis-Hastings is being investigated.

CC151 Room Aula H BAYESIAN STATISTICS **Chair: Leonardo Egidi**

C0405: **Testing normalizing flows for posteriors in variational Bayes**
Presenter: **Tim Kutzker**, Humboldt University Berlin, Germany
Co-authors: Nadja Klein

Normalizing flows (NFs) model complex probability density functions as concatenations of invertible (backward) transformations of simple densities with sound statistical characteristics. Data generation in return requires forward transformations. The ability to approximate and sample from arbitrary densities sufficiently well also brought NFs more and more into the spotlight to serve as variational densities in approximating complex posterior distributions through variational Bayes. To be practical in applications and to compete with existing methods such as MCMC, however, NFs must be sufficiently fast and efficient in both the forward and backward direction, which usually oppose each other. We propose a statistical test that first ensures that the NF approximates the probability density function sufficiently well and second follows the principle of parsimony ensuring the data is generated as fast as possible. For this purpose, we scale the multidimensional (density) test problem to the univariate (two-sample) Kolmogorov-Smirnov test approach by considering a standard uniformly distributed transformation of the highest probability density region, while retaining computational efficiency. We highlight the merits of our tests in a detailed MC study and a real data example.

C0422: **Bayesian estimation versus maximum likelihood estimation in the Weibull-power law process**
Presenter: **Alicja Jokiel-Rokita**, Wroclaw University of Science and Technology, Poland
Co-authors: Ryszard Magiera

The Bayesian approach is applied to the estimation of the Weibull-power law process (WPLP) parameters as an alternative to the maximum likelihood (ML) method in the case when the number of events is small. For the process model considered, we propose to apply the independent Jeffreys prior distribution and we argue that this is a useful choice. Comparisons were also made between the accuracy of the estimators obtained and those obtained by using other priors – informative and weakly informative. The investigations show that the Bayesian approach in many cases of a fairly broad collection of WPLP models can lead to the Bayes estimators that are more accurate than the corresponding ML ones when the number of events is small. The problem of fitting the WPLP models, based on ML and Bayes estimators, to some real data is also considered. It is shown that the TTT-concept, used in the reliability theory, is not fully useful for the WPLP models, and it may be so for some other trend-renewal processes. In order to assess the accuracy of the fitting to the real data considered, two other graphical methods are introduced.

C0437: **Backward filtering forward guiding for Markov processes**
Presenter: **Frank van der Meulen**, Delft University of Technology, Netherlands
Co-authors: Moritz Schauer

Consider a Markovian process X that evolves on a tree where transitions over edges correspond to running a continuous-time Markov process for a fixed time interval. At each vertex, leaves can be attached that represent observations. A key example consists of a diffusion process on a tree, appearing for example in phylogenetics. Assume the forward dynamics are parametrised by the parameter theta. We will discuss the Backward Filtering Forward Guiding algorithm for sampling X conditional on its values at the leaf vertices. This in turn can be exploited for designing Bayesian computational methods such as MCMC and SMC for inferring theta.

C0448: **A Bayesian nonparametric estimation of entropy for circular data**
Presenter: **Najmeh Nakhaeirad**, University of Pretoria, South Africa
Co-authors: Andriette Bekker, Mohammad Arashi, Sollie Millard

Entropy is a widely-used information theoretic measure; however, the major problem in information theoretic analysis of data is the reliable estimation of entropy, especially from small samples. Furthermore, there is a gap in the literature regarding the estimating of entropy for circular data. Circular data comes from several domains with special emphasis on the phases of periodic phenomena and directions such as biology, physics, neuroscience, earth sciences, economics and meteorology. A Bayesian approach is implemented to obtain the nonparametric estimation of Shannon entropy for circular data. Three different estimators are proposed and their performance is compared via a simulation study. We close with the application of Shannon entropy in circular data analysis.

C0650: **Exact likelihood for inverse gamma stochastic volatility models**
Presenter: **Blessings Majoni**, National graduate institute for policy studies, Japan
Co-authors: Roberto Leon-Gonzalez

A novel closed form expression of the likelihood for the inverse gamma stochastic volatility model is obtained. It is shown that by marginalizing out the volatilities the model that we obtain has the resemblance of a GARCH in the sense that the formulas that we get are similar, which simplifies computations significantly. We also obtain methods to draw the latent volatilities directly from their posterior distributions. Recent literature has also attempted to obtain closed-form solutions for the likelihood in stochastic volatility models. However, the literature has only obtained such a solution for non-stationary models, or for the stationary Gamma Stochastic Volatility model. We compare the empirical fit of our proposed model with the previous literature.

CC215 Room Aula I CLASSIFICATION **Chair: Marialuisa Restaino**

C0406: **Multi-class classification with imbalanced data: The choice of a categorical classifier**
Presenter: **Silvia Golia**, University of Brescia, Italy
Co-authors: Maurizio Carpita

The issue of the choice of the categorical classifier is discussed, that is the procedure that, starting from the probabilities assigned to all the categories by a suitable method (probabilistic classifier), transforms these probabilities into a single class. The focus is on multi-class target variables, that is, variables that admit k non-overlapping classes and the units are to be classified into one, and only one, of them. The standard choice is the Bayes Classifier (BC), which assigns, based on the probabilistic classifier, a unit to the most likely class. Nevertheless, BC has some limits with rare classes, given that it favors the prevalent class, and in situations in which there is not a class of interest or it is not prevalent, the BC cannot be the best choice. The aim is to investigate, through an extensive simulation study, the classification performances of the BC versus two alternatives, that is the Max Difference Classifier (MDC) and Max Ratio Classifier (MRC). The obtained results show that, in terms of Macro Recall and F-score measures and stability in the face of increasing class imbalance, MDC and MRC are better alternatives to BC. Some real case studies confirm what is observed in the simulation.

C0419: **A dimensionality expansion methodology for loss optimization in cost sensitive problems**
Presenter: **Jorge C-Rella**, Abanca servicios financieros, Spain
Co-authors: Ricardo Cao, Juan Vilar Fernandez

In all the current cost-sensitive classification models there is an intrinsic loss of information due to a blur in the consideration of the exogenous variable that defines gains/losses. The most extended approaches consider decision rules based on an estimated fraud probability, considering a cost matrix. To tackle this blur, two refinements are introduced. First, a loss function is considered a performance metric, which provides a more realistic measure for the expected results in practice. Second, a new decision space is constructed considering the estimated probability and the exogenous variable influencing the loss. This space permits a more flexible search for the decision region. An algorithm (2-DDR) is proposed to optimize the loss function over the decision space with plenty of freedom. The estimated decision region includes, as particular cases, classical approaches. As a consequence, an improvement is always achieved. This is tested and assessed with a wide range of simulations with varying difficulty and structure, showing the algorithm robustness and the systematic improvement with respect to previous approaches.

C0483: **Doping control analysis in athletes steroid profile: A multivariate Bayesian learning approach**
Presenter: **Dimitra Eleftheriou**, University of Glasgow, United Kingdom

Anabolic androgenic steroids (AAS) are frequently detected as doping substances in competitive sports. In order to detect AAS doping with pseudo-endogenous steroids, i.e. steroids that are produced in the human body like testosterone, urinary concentrations of the athlete's steroid profile are measured over time in the steroidal module of the Athlete Biological Passport (ABP). This research work focuses on extending the current univariate Bayesian model, which monitors each biomarker in the steroid profile separately, to a multivariate multilevel adaptive model, which is able to accommodate repeated measurements from various sensitive biomarkers and their concentration ratios. The developed methodology was applied to urine sample data obtained from professional athletes. Among these samples, normal, atypical, and abnormal values were identified. An anomaly detection technique based on a one-class classification (OCC) algorithm was carried out to detect anomalies within the athletes' steroid profiles, either due to AAS misuse or other confounding factors. In a Bayesian context, the main idea is to construct adaptive decision boundaries around normal concentration values as new recordings come, and differentiate them from the abnormal ones. Improved prediction performance was obtained compared to standard methodologies suggesting the proposed approach as an improved tool for doping detection.

C0652: **Fuzzy classification with distance-based prototypes**
Presenter: **Itziar Irigoien**, University Basque Country, Spain
Co-authors: Concepcion Arenas

Supervised or unsupervised classification of objects are important areas of research and needed in practical applications in a variety of fields such as environmental sciences, medicine, economy and psychology. Distance-based approaches offer a complementary perspective to classical units×variables techniques. Besides fuzzy approaches bring the opportunity to handle situations where there is not a clear-cut relationship between units and where units present different degrees of membership. The aim is to combine both aspects and introduce a novel Fuzzy Classification method. To that end, first, fuzzy versions of distance-based concepts such as geometric variability, proximity function, distance between classes and depth function are extended. Then, inspired by previous works, a fuzzy classification methodology is proposed including the aforementioned distance-based perspective. The proposed methodology covers supervised and non-supervised tasks and in contrast with more classical approaches, offers characteristic prototypes of a given data set instead of centroids. To show its effectiveness, the proposed approach was compared with Supervised Fuzzy Partitioning on some classical datasets as well as simulated datasets. Finally, the results we obtained on real datasets are reported showing the good performance of the new methodology.

C0594: **Discriminant analysis with corrupted label data using subject similarity**
Presenter: **Masaaki Okabe**, Doshisha University, Japan
Co-authors: Hiroshi Yadohisa

In the classification task, the labels of the obtained training data are assumed to be correct. However, the data labeled by humans and the labels assigned to objects may be incorrect due to problems such as mislabeling. If the labels in the training data are incorrect, the classification accuracy of the discriminant model may be reduced. The previous study assumes that given a true label, features and corrupted occurrences are independent of one another. In other words, they assume that mislabeling occurs randomly. In this situation, when the balanced error rate (BER) is used as the objective function, it is shown that the discriminant model that optimizes the objective function for data with corrupted labels optimizes the BER for data without corrupted labels. However, the classification may not work well when the label corruptness is correlated with the features. For example, if a label error depends on a feature, the label error will be correlated with the feature. The aim is to solve this problem by treating corrupted labels and weighting objects with features.

CC159 Room Aula E **ALGORITHMS AND COMPUTATIONAL METHODS** **Chair: Bettina Gruen**

C0302: **Exact simulation of continuous max-id processes**
Presenter: **Florian Brueck**, Technical University Munich, Germany

An algorithm for the unbiased simulation of max-(or min-)id stochastic processes is presented. The algorithm only requires the simulation of finite Poisson random measures and avoids the necessity of computing conditional distributions of infinite (exponent)measures. Special emphasis is put on the simulation of exchangeable max-(or min-)id sequences, in particular exchangeable Sato-frailty sequences.

C0537: **Tricks that accelerate matrix multiplication on CPUs**
Presenter: **Martin Schlather**, Universitat Mannheim, Germany
Co-authors: Alexander Freudenberg

General approaches to accelerate matrix multiplication are parallel computing, SIMD programming and the Strassen algorithm. It is surprising that, although SNP matrices are large, the Strassen algorithm might be significantly outperformed by fine-tuned standard algorithms, when calculating the genomic relationship matrix. We give an introduction to low-level matrix multiplication and show the acceleration when SIMD commands, cache and registers are used exhaustively. Clearly, the results highly depend on the compression level.

C0573: **GPU routines for accelerated genomic calculations**
Presenter: **Alexander Freudenberg**, University of Mannheim, Germany
Co-authors: Martin Schlather

Genomic datasets used in empirical research are steadily growing in size and advances in computing power can only partially offset the associated computational demand. A suboptimal utilization of resources can lead to significant increases in hardware requirements or computation times. We explore the benefits of highly finetuned GPU routines for the calculation of important population statistics, which utilize instruction sets of modern NVIDIA hardware. We showcase examples as part of our implementation in the R package miraculix which is intended to be accessible to a broad range of practitioners.

C0605: **Extending linear programming ecological inference methods by machine learning**
Presenter: **Jose M Pavia**, Universitat de Valencia, Spain

Ecological inference (EI) is devised to forecast unknown inner-cells of two-way contingency tables by inferring conditional distribution probabilities. This outlines one of the more conspicuous and long-standing social science problems present in many areas, with political science and sociology being the disciplines where they are chiefly more frequent. For instance, EI algorithms are used to estimate vote transfer matrices between elections, infer split-ticket voting behaviors or reveal social and racial voting patterns. In the last years, we have experienced an explosion of methods to solve these problems from the Bayesian approach, based on a hierarchical multinomial-Dirichlet Bayesian model. The use of these methods, however, requires highly trained analysts and usually entails high computational costs. Recently, a new family of algorithms that considerably simplify the resolution of these problems has been proposed based on mathematical programming. The first wave of these new algorithms, which are at least as accurate as the Bayesian-based algorithms, are available in the R-package lphom. The goal is to show the accuracy improvements that we are achieving by integrating statistical learning approaches in this new methodology. Specifically, through the use of new algorithms based on (inspired by) bagging, genetic boosting and reinforcement learning.

C0657: **Fussed nearly-isotonic signal approximation**

Presenter: **Vladimir Pastukhov**, Chalmers, Sweden

The aim is to estimate sparse piecewise monotone signals and introduce fussed nearly-isotonic signal approximation. We provide different numerical solutions to the optimisation problem, show that it is computationally feasible, derive the degrees of freedom for the estimator and study the asymptotic properties. Based on the simulations we show that the estimator behaves well and in certain cases, it outperforms other constrained estimators.

CC220 Room Aula F COMPUTATIONAL AND FINANCIAL ECONOMETRICS II **Chair: Niklas Ahlgren**

C0588: **Adaptive estimation for somenonparametric instrumental variablemodels with full independence**

Presenter: **Fabian Dunker**, University of Canterbury, New Zealand

The problem of endogeneity in statistics and econometrics is often handled by introducing instrumental variables (IV) which fulfil the mean independence assumption, i.e. the unobservable is mean independent of the instruments. When full independence of IVs and the unobservable is assumed, nonparametric IV regression models and nonparametric demand models lead to nonlinear integral equations with unknown integral kernels. We prove convergence rates for the mean integrated square error of the iteratively regularized Newton method applied to these problems. Compared to related results we derive stronger convergence results that rely on weaker nonlinearity restrictions. We demonstrate in numerical simulations for a nonparametric IV regression that the method produces better results than the standard model.

C0282: **Generalized autoregressive conditional betas**

Presenter: **Francesco Violante**, ENSAE ParisTech, France

Co-authors: Stefano Grassi

A new class of multivariate models is introduced allowing for observation-driven time-varying slope coefficients in a system of conditionally heteroskedastic simultaneous multiple regressions. These processes, dubbed Generalized Conditional Autoregressive Beta (GCAB), introduce a structural layer tailored to the linear asset pricing framework, solving the problem of incorporating time variation in the exposure of assets to risk factors in the asset pricing equation. This class of models accommodate for large cross-sectional dimensions (both in terms of regressors and regressands), and allow parametric cross-sectional restrictions, which are key for validation of asset pricing models. The proposed dynamics naturally accommodate the coexistence of constant and time-varying betas (that can be validated via testable hypotheses), and introduce new economically meaningful mechanisms of propagation of shocks, tagged beta spillovers. We derive stationarity and uniform invertibility conditions and, to mitigate the problem of parameter proliferation in large dimensions, we also provide beta and covariance tracking constraints. We propose a variety of computationally convenient (parallel and sequential) quasi maximum likelihood estimators, and we investigate their finite sample properties by means of Monte Carlo experiments. Finally, the GCAB is used to illustrate the role of beta spillovers in the Fama-French three factors asset pricing framework.

C0519: **A volatility model with a time-varying intercept**

Presenter: **Niklas Ahlgren**, Hanken School of Economics, Finland

Co-authors: Alexander Back, Timo Terasvirta

A GARCH model augmented by a time-varying intercept is proposed. The intercept is parameterised by a logistic transition function with rescaled time as the transition variable. This formulation provides a simple and flexible way to capture deterministic non-linear changes in the conditional and unconditional variances. It is common for financial time series to exhibit these types of shifts. By making the intercept a smooth function of time, it is possible to capture changes that occur gradually, rather than abruptly as in regime switching models. The model is globally nonstationary but locally stationary. We use the theory of locally stationary processes to derive the asymptotic properties of the quasi-maximum likelihood estimator (QMLE) of the parameters of the model. We show that the QMLE is consistent and asymptotically normally distributed. To corroborate the results, we provide a small simulation study. An empirical application to Intel Corporation stock returns demonstrates the usefulness of the model. We find that the persistence implied by the standard GARCH model parameter estimates is reduced by incorporating a time-varying intercept. In particular, estimates that suggest an integrated volatility model are reduced to lie within the stationary region.

C0669: **Estimating crypto market betas**

Presenter: **Jan Sila**, UTIA AV CR, v.v.i., Czech Republic

Co-authors: Ladislav Kristoufek

The seminal financial problem of estimating market betas for crypto assets is investigated. We present empirical evidence of limited predictability of the future OLS betas, the theoretically optimal hedge against market risk. Our main finding is that non-OLS procedures better predict future realized OLS betas. This result correlates with recent stock market literature, but the general stability of crypto betas is much lower. Our work has implications for modern asset pricing theories measuring exposures to risk factors. The results also show a strong dependence on the selection of the market index, where we compare the ones recently used in the literature and the industry.

C0150: **Discriminating direct from induced equilibrium-mean shifts**

Presenter: **David Hendry**, University of Oxford, United Kingdom

Co-authors: Jennifer L Castle, Jurgen Doornik

Equilibrium-mean, or location, shifts can result directly from changes in intercepts with constant dynamics, or be induced by shifts in dynamics (or other parameters) when data means are non-zero. The impacts of in-sample induced shifts substantively modify previous taxonomies of forecast errors. Step-indicator saturation helps detect any resulting location shifts. However, even when all relevant variables in the data generation process (DGP) and all indicators matching DGP shifts are selected in the forecasting model, mis-forecasting can occur. To discriminate direct from induced shifts, we add to the model multiplicative indicators formed by interacting all selected step indicators with the lagged regressand. When equilibrium-mean or location shifts are induced by changes in dynamics, forecasts can be markedly improved when these interactive indicators are included.

Tuesday 23.08.2022 14:15 - 15:45 Parallel Session C – COMPSTAT2022

CI007 Room Aula Q BOOTSTRAP AND RESAMPLING IN CLUSTER ANALYSIS Chair: Christian Hennig

C0394: **Validation of cluster analysis results on validation data: A systematic framework**

Presenter: **Theresa Ullmann**, LMU Munich, Germany

Co-authors: Christian Hennig, Anne-Laure Boulesteix

Cluster analysis refers to a wide range of data analytic techniques for class discovery and is popular in many application fields. To assess the quality of a clustering result, different cluster validation procedures have been proposed in the literature. While there is extensive work on classical validation techniques, such as internal and external validation, less attention has been given to validating and replicating a clustering result using a validation dataset. Such a dataset may be part of the original dataset, which is separated before analysis begins, or it could be an independently collected dataset. We present a systematic, structured review of the existing literature on this topic. For this purpose, we outline a formal framework that covers most existing approaches for validating clustering results on validation data. In particular, we review classical validation techniques such as internal and external validation, stability analysis, and visual validation, and show how they can be interpreted in terms of our framework. We define and formalize different types of validation of clustering results on a validation dataset, and give examples of how clustering studies from the applied literature that used a validation dataset can be seen as instances of our framework.

C0596: **Resampling methods for exploring cluster stability**

Presenter: **Friedrich Leisch**, Universitaet fuer Bodenkultur Vienna, Austria

Model diagnostic for cluster analysis is still a developing field because of its exploratory nature. Numerous indices have been proposed in the literature to evaluate goodness-of-fit, but no clear winner that works in all situations has been found yet. Derivation of (asymptotic) distribution properties is not possible in most cases. Resampling schemes provide an elegant framework to computationally derive the distribution of interesting quantities describing the quality of a partition. Special emphasis will be given to the stability of a partition, i.e., given a new sample from the same population, how likely is it to obtain a similar clustering? Another application is to systematically compare partitions with a growing number of clusters to assess which clusters are stable and found repeatedly even when the number of clusters increases.

C0430: **Clustering of single cell RNAseq data: An integrated analysis using multiple methods and robust clustering solutions**

Presenter: **Mohamad Zafer Merhi**, Hasselt University, Belgium

Co-authors: Ziv Shkedy, Ahmed Essaghir, Dan Lin

Clustering single cell RNA-seq data is a central step in the identification of cell types in single cell RNA-seq data experiments. Through the clustering unsupervised analysis, we are able to find groups of cells based on similarities in their expression profiles which allows us to associate subsets of cells (belong to the same cluster) with a biological pathway. Despite recent advancements in clustering tools and methods aimed at clustering single-cell RNA-seq data, many challenges and factors still need to be investigated. For example, a collection of clustering methods applied to the same single cell RNA-seq data often results in a variety of clustering solutions. Even in the case that a single clustering method is used, a change in the parameter settings typically produces a different clustering solution. In the current study, we assess the performance of selected clustering methods and focus on the similarity between the clustering solutions obtained for the different methods. We discuss the methodology to identify a robust clustering solution for a given single cell RNA-seq data and present diagnostic plots to investigate, for a given method, the influence of the parameter setting on the solution. All methods are applied to real-life (and publicly available) single cell RNAseq data. Software tools to conduct the proposed analysis are presented (and publicly available) as well.

CO115 Room Virtual Room R1 LATENT VARIABLE AND PSYCHOMETRIC MODELLING (VIRTUAL) Chair: Michela Battauz

C0210: **Equating tests with mixed format tests**

Presenter: **Marie Wiberg**, Umea University, Sweden

An achievement test typically contains items with different scoring formats, such as dichotomously scored and polytomously scored items. The aim is to propose a novel approach to equate achievement tests containing dichotomously scored and polytomously scored items. Previous research has focused on item response theory (IRT), observed-score equating, IRT true-score equating, frequency estimation and chain equating. We present a novel approach which can be used to equate two test versions which have a mixed item format. We focus on the equivalent group design and illustrate the method with real test data from a national test in mathematics. We also examined different conditions in a simulation study, including proportions of binary and polytomous items and when a different level of ability is seen in the groups which receive the different test versions. The proposed approach shows stable equating results and appears to be a good alternative to commonly used equating methods for mixed item format.

C0211: **Monitoring the Brunelleschi's Dome through latent variable models**

Presenter: **Silvia Bacci**, Department of Statistics Computer Science Applications University of Florence, Italy

Co-authors: Bruno Bertaccini, Fabrizio Cipollini

Brunelleschi's Dome of Santa Maria del Fiore, in Florence (IT), is one of the most famous symbols of the Reinassance in the World. Monitoring its stability and detecting any atypical behavior is a priority in protecting this important monument. The first cracks in the Dome appeared at the end of the 15th century, and nowadays, they are present in all the Dome's webs, although with a heterogeneous distribution. A monitoring system has been installed in the Dome since 1955 to monitor the behavior of the cracks; today, it counts more than 160 instruments, such as mechanical and electronic deformometers, thermometers, piezometers. The analyses carried out to date show slight increases in the size of the main cracks and, at the same time, a clear relationship with some environmental variables. However, due to the extension of the monitoring system and the complexity of collected data, an analysis that involves all the variables detected has never been presented in any of the studies conducted in the past. We aim to formulate and estimate a latent variable model to find out simplified structures (i.e., latent common factors) that summarize the measurements coming from the different instruments and explain the overall behavior of the Dome across time.

C0451: **Psycho COVID-19: Evaluating the risk of the psycho-physical impact of the pandemic**

Presenter: **Roberto Di Mari**, Universita' di Catania, Dipartimento di Economia e Impresa, Italy

This is a systematic study of the psychological consequences of COVID-19, based on a large sample of online answers to a structured questionnaire of 20 questions. The questionnaire has been designed by a group of well-known psychologists and physicians coordinated by Dr. Vito Tummino, and has been implemented online by the Provincial Health Authority of Ragusa, Italy. Using techniques of MCDA (Multiple Criteria Decision Analysis), we are able to exploit the opportunity of interaction between the statistician and a clinical expert, in order to obtain an aggregate distribution-free score for each class of items. The resulting two scores, each on a 0-1 continuous scale, represent, respectively, physical and psychological risk of experiencing maladjustment and stress disorder. Subsequently, we specify a novel 0-1 inflated semisupervised ordinal latent class model to build a classifier, which, by leveraging on all available information from the two scores, groups units into 9 "overall risk" classes - ranked from 1 (no risk at all or negligible risk) to 9 (very high or maximum risk). The individual posterior membership probabilities are then elaborated to be used i) by potential patients as a tool for raising self-awareness, and ii) by aiding professionals as an individual and aggregate monitoring tool.

C0634: **A stochastic optimization algorithm for pairwise likelihood estimation of factor models with ordinal data**

Presenter: **Giuseppe Alfonzetti**, University of Padova, Italy

The typical computational challenge of maximum likelihood estimation for non-linear latent variable models is the integration of the latent variables from their joint likelihood with the observed data. When tackled with a quadrature approach, the integration implies an exponential complexity in the dimension of the latent space. Moving to a pairwise likelihood estimator allows replacing this integral with the sum of many bidimensional problems. Unfortunately, their amount grows with the square of the number of items, which prevents the estimation to be feasible on large datasets. To solve this problem in the common case of ordinal data, a stochastic optimization algorithm is proposed in order to scale the estimation on large factor models. At each iteration, a cheap approximation to the gradient of the pairwise likelihood is computed by sampling a small subset from the complete pool of pairs. The complexity per iteration achieved does not depend on the sample size, grows only quadratically with the dimension of the latent space and potentially only linearly with the number of items.

CO105 Room Aula B ISBA SESSION: APPLIED COMPUTATIONAL BAYES (VIRTUAL) Chair: Giacomo Zanella

C0222: **Scalable inference for epidemic models with individual level data**
Presenter: **Panayiota Touloupou**, University of Birmingham, United Kingdom
Co-authors: Simon Spencer, Barbel Finkenstadt

As individual level epidemiological and pathogen genetic data become available in ever increasing quantities, the task of analysing such data becomes more and more challenging. Inferences for this type of data are complicated by the fact that the data is usually incomplete, in the sense that the times of acquiring and clearing infection are not directly observed, making the evaluation of the model likelihood intractable. A solution to this problem can be given in the Bayesian framework with unobserved data being imputed within Markov chain Monte Carlo (MCMC) algorithms at the cost of considerable extra computational effort. Motivated by this demand, we describe a novel method for updating individual level infection states within MCMC algorithms that respects the dependence structure inherent within epidemic data. We apply our new methodology to an epidemic of Escherichia coli O157:H7 in feedlot cattle in which eight competing strains were identified using genetic typing methods. We show that surprisingly little genetic data is needed to produce a probabilistic reconstruction of the epidemic trajectories, despite some possibility of misclassification in the genetic typing. We believe that this complex model, capturing the interactions between strains, would not have been able to be fitted using existing methodologies.

C0310: **Concentration and robustness of discrepancy-based ABC through Rademacher complexity**
Presenter: **Sirio Legramanti**, University of Bergamo, Italy
Co-authors: Daniele Durante, Pierre Alquier

Approximate Bayesian Computation (ABC) typically employs summary statistics to measure the discrepancy among the observed data and synthetic data generated from each proposed parameter value. However, finding good summary statistics (that are close to sufficiency) is non-trivial for most of the models for which ABC is needed. This motivated summary-free versions of ABC based on discrepancies between the empirical distributions of observed and synthetic data. The studies on the properties of the corresponding ABC posteriors are quite uneven, ranging from empirical assessments to more theoretical investigations. Even when available, the existing theory is often limited to a single discrepancy or relies on hypotheses that are difficult to verify. We propose a unifying view through Rademacher complexity over a general class of discrepancies known as integral probability semimetrics, which include the maximum mean discrepancy and the total variation, Kolmogorov-Smirnov, and Wasserstein distances. For rejection ABC based on this class of semimetrics, we prove results on both posterior concentration and robustness. Such results connect the properties of the ABC posterior to the Rademacher complexity of the class of test functions that characterizes each integral probability semimetric. This provides a new understanding of why some discrepancies work well with ABC and others do not.

C0480: **Spatially-varying Bayesian predictive synthesis for flexible and interpretable spatial prediction**
Presenter: **Kenichiro McAlinn**, Temple University, United States
Co-authors: Masahiro Kato, Shonosuke Sugasawa, Kosaku Takanashi, Danielle Cabel

Spatial data are characterized by their spatial dependence, which is often complex, non-linear, and difficult to capture with a single model. Significant levels of model uncertainty – arising from these characteristics – cannot be resolved by model selection or simple ensemble methods, as performances are not homogeneous. We address this issue by proposing a novel methodology that captures spatially-varying model uncertainty, which we call spatial Bayesian predictive synthesis. Our proposal is defined by specifying a latent factor spatially-varying coefficient model as the synthesis function, which enables model coefficients to vary over the region to achieve flexible spatial model ensembling. Two MCMC strategies are implemented for full uncertainty quantification, as well as a variational inference strategy for fast point inference. We also extend the estimations strategy for general responses. A finite sample theoretical guarantee is given for the predictive performance of our methodology, showing that the predictions are exact minimax. Through simulation examples and two real data applications, we demonstrate that our proposed spatial Bayesian predictive synthesis outperforms standard spatial models and advanced machine learning methods, in terms of predictive accuracy, while maintaining interpretability of the prediction mechanism.

C0598: **Evidence approximation and Bayesian model choice**
Presenter: **Christian Robert**, Universite Paris-Dauphine, France

Evidence approximation is a central object of Bayesian inference and despite numerous advances in the past decades, there still remain challenges to be met, especially when the sample size is large. We review here some robust solutions like the reverse logistic regression and a modified harmonic mean estimator, before proposing a related algorithm for Bayesian model choice.

CO017 Room Aula C ANALYSIS OF RANKING DATA Chair: Philip Yu

C0350: **Empirical Bayes on a shoestring and other applications**
Presenter: **Mayer Alvo**, University of Ottawa, Canada

The Pearson system of distributions is exploited to obtain expressions for the posterior mean and posterior variance in Tweedie's formula. This has interesting consequences in the study of micro-array data where the interest is uncovering unusual genes. Another application deals with the development of new non-parametric tests for the two sample problems of location and of scale. These tests are based on an approximation of the locally optimal score test statistics.

C0290: **A computationally efficient non-parametric signal estimation approach for ranking data**
Presenter: **Michael Georg Schimek**, Medical University of Graz, Austria
Co-authors: Luca Vitale, Bastian Pfeifer, Michele La Rocca

The ranking of items is widely used to rate their relative quality or relevance across multiple lists of assessments. Typically, the list length p is in the thousands and the number of lists $n << p$. Our interest, beyond classical rank aggregation, is to estimate the, usually unobservable, latent signals that inform a consensus ranking. Under the only assumption of independent assessments, we introduce indirect inference via convex optimisation in combination with computationally efficient Poisson Bootstrap. The mathematical formulation of the signal estimation problem is based on pairwise comparisons of all items with respect to their rank positions. The order relations are represented by a system of inequalities for optimisation. The transitivity property of rank scales allows us to reduce substantially the number of constraints associated with the full set of item comparisons. The key idea is the successive reduction of the ranker-induced errors until optimal latent signal estimates are obtained. Its advantage is a substantial reduction in the computational burden and the possibility to handle n«p data problems. The power of this novel approach is demonstrated in a large bio-medical data problem.

C0338: **Ensemble methods for item-weighted label ranking: A comparison**
Presenter: **Mariangela Sciandra**, Universita degli studi di palermo, Italy
Co-authors: Antonella Plaia, Alessandro Albano

Label Ranking (LR), an emerging non-standard supervised classification problem, aims at training preference models that order a finite set of labels based on a set of predictor features. Traditional LR models regard all labels as equally important. However, in many cases, failing to predict the ranking position of a highly relevant label can be considered more severe than failing to predict a trivial one. Moreover, an efficient LR classifier should be able to take into account the similarity between the items to be ranked. Indeed, swapping two similar elements should be less penalized than swapping two dissimilar ones. The contribution is to formulate more flexible item-weighted label ranking models that make use of well-known decision tree ensemble models; respectively: bagging, random forest and boosting. The three proposed weighted LR classifiers encode the similarity structure and the individual label importance provided by a domain expert. The predictive performances of the three algorithms are compared, through simulations, to determine which ensemble procedure produces the best results for different noise levels and weight sets.

C0304: **Social order statistics models for ranking data**
Presenter: **Philip Yu**, The Education University of Hong Kong, Hong Kong
Co-authors: Jiaqi Gu

Human interaction and communication have become essential features of social life. Individuals' preferences may be influenced strongly by those of their peers or friends in a social network. So far, traditional ranking models do not account for such social network dependency. We introduce a new class of models called social order statistics (SOS) models to learn ranking data in social networks. The new models combine the order statistics models and spatial autoregressive models to account for social dependencies among the individuals. A flexible formulation of weight matrices in the spatial model is adopted to provide diverse network effects among the individuals for different items. Efficient and scalable MCMC algorithms are developed to perform Bayesian inference in a parallel manner for large networks with even a few thousand nodes. Simulation and empirical studies demonstrate the usefulness of our proposed inference procedures and reveal that social network effects could be different for individuals' preferences towards different items in a social relationship.

CO033 Room Aula D SOME ADVANCES IN MULTIVARIATE AND FUNCTIONAL STATISTICS Chair: Enea Bongiorno

C0237: **Functional data clustering with outlier detection**
Presenter: **Julien Jacques**, Universite de Lyon, France
Co-authors: Martial Amovin, Irene Gannaz

With the emergence of numerical sensors in many aspects of everyday life, there is an increasing need to analyze high-frequency data, which can be seen as discrete observations of functional data. The focus will be on the clustering of such functional data in order to ease their modeling and understanding. To this end, a novel clustering technique for multivariate functional data is presented. This method is based on a functional latent mixture model, which fits the data in group-specific functional subspaces through a multivariate functional principal component analysis. In such clustering analysis, the presence of outliers can confuse the notion of cluster. Consequently, a contaminated version of the previous mixture model is proposed. This model both clusters the multivariate functional data into homogeneous groups and detects outliers. The main advantage of this procedure over its competitors is that it does not require us to specify the proportion of outliers. The model inference is performed through an Expectation-Conditional Maximization algorithm, and the BIC criterion is used to select the number of clusters. Numerical experiments on simulated data demonstrate the high performance achieved by the inference algorithm. In particular, the proposed model outperforms competitors. Its application on the real data, which motivated this study allows us to detect abnormal behaviors correctly.

C0323: **Statistical depth for multivariate and functional data: Recent progress and perspectives**
Presenter: **Stanislav Nagy**, Charles University, Czech Republic

Statistical depth is a tool of nonparametric analysis that generalises quantiles, rankings, and orderings to multivariate and non-Euclidean data. While a rich body of literature on various depths and depth-like procedures exists, many open problems still stimulate research in the area. We draw interesting connections of depth with topics firmly established in convex geometry and statistical machine learning. These observations allow us to resolve several open problems of statistical depth.

C0384: **The two sample problem for functional data**
Presenter: **Manuel Febrero-Bande**, University of Santiago de Compostela, Spain
Co-authors: Wenceslao Gonzalez-Manteiga, Ana Colubi, Gil Gonzalez-Rodriguez

The aim is to present a new test for the two-sample problem, i.e. when it is reasonable to conclude that two samples (possibly with different sizes) belong to the same distribution. The two-sample problem is solved in univariate samples through the use of the distribution functions that are not readily available in functional data scenarios. The proposal is based on energy statistics and can be applied (calibrated) to all scenarios including functional data analysis. Previous approaches in this context will be compared against the new proposal through simulation studies.

C0305: **Customizing the dimensionality of functional data**
Presenter: **Aldo Goia**, Universita' del Piemonte Orientale, Italy
Co-authors: Enea Bongiorno

The representation of functional data in a small dimension is a very important task. Usually, it is performed by using the well-known truncated Karhunen-Loeve expansion where the truncation threshold is selected suitably: it is a global method since once the dimension is chosen it is used for all the curves. Nevertheless, this approach is not optimal in the sense that some curves could be represented in a lower dimension and for other ones a larger dimension would be desirable. To obtain a more parsimonious representation of the data, a local dimensionality reduction method can be used: it is based on a nonparametric estimate of the correction term appearing in a Small ball probability factorization for functional Hilbert data. The latter can be interpreted as a measure of the quality of the representation of functional data in small dimensions and its theoretical properties are investigated. To assess the ability of the local selection approach, some applications to simulated and real datasets are performed.

CO125 Room Aula I STATISTICAL ANALYSIS OF NETWORKS: APPLICATIONS IN CYBER-SECURITY Chair: Francesco Sanna Passino

C0280: **Computer network security datasets**
Presenter: **Kate Highnam**, Imperial College London, United Kingdom

As cyber threats continue to advance, new defences utilise intelligent statistical solutions. However, the required domain expertise can prevent statisticians from applying their methods to cyber security applications. To bridge this gap, we present our publicly available datasets collected from honeypots, intentionally vulnerable systems exposed to the Internet to observe real-world attacks. Our honeypots recorded millions of data points from internal host processes and network traffic, containing highly structured but heterogeneous features. The dataset also includes simultaneous logging of multiple identical systems, where only some were exploited by adversaries, for control comparison. By deploying in environments that limit the noise in the data, we enable non-security experts to demonstrate their methods ability against real adversaries in real systems. We will describe our honeypots and their data, compare with other publicly available datasets, and discuss important research questions in network security.

C0365: **Unsupervised attack pattern detection in cyber-security using topic modelling**
Presenter: **Anastasia Mantziou**, Imperial College London, United Kingdom
Co-authors: Francesco Sanna Passino, Nick Heard, Philip Thiede, Ross Bevington

Cyber systems are constantly under threat of intrusion attempts. Attacks are usually carried out with one underlying specific intent, or from groups of actors with similar objectives. Therefore, discovering such patterns is extremely valuable to threat experts. From a statistical point of view, this objective translates into a clustering task. The aim is to explore topic models for clustering session data collected on honeypots, particular hosts designed to entice malicious intruders. The main practical implications of clustering the sessions are two-fold: finding similar groups and identifying outliers. An array of methodologies is considered, suitably adapted to the challenges encountered with computer network data. In particular, the concepts of primary topics, session-level and command-level topics are introduced, along with a secondary topic for instruction representing common high-frequency commands. Furthermore, the proposed method is extended to allow for an unbounded number of latent intents. The methodologies are used to discover an unusual MIRAI variant which attempts to take over existing coin miner infrastructure.

C0379: **Spectral embedding of weighted graphs**
Presenter: **Ian Gallagher**, University of Bristol, United Kingdom
Co-authors: Patrick Rubin-delanchy, Carey Priebe, Andrew Jones, Anna Bertiger

The statistical analysis of a weighted graph through spectral embedding is considered. Under a latent position model in which the expected adjacency matrix has a low rank, we prove uniform consistency and a central limit theorem for the embedded nodes, treated as latent position estimates. In the special case of a weighted stochastic block model, this result implies that the embedding follows a Gaussian mixture model with each component representing a community. We exploit this to formally evaluate different weight representations of the graph using Chernoff information. For example, in a network anomaly detection problem where we observe a p-value on each edge, we recommend against directly embedding the matrix of p-values, and instead, using threshold or log p-values, depending on network sparsity and signal strength.

C0445: **Peer-group behavior analytics modelling with mutually exciting point process graphs**
Presenter: **Henrique Helfer Hoeltgebaum**, Securonix, United Kingdom
Co-authors: Francesco Sanna Passino

The increasingly complex threat technology adopted by malicious entities to evade existing defences in cyber environments is a growing concern for society. Cyber-security analysts have difficulties coping with the increasingly large number of alerts received on any given day. This is mainly due to the low precision of existing detectors, which end up producing a substantial number of false positives. Usually, several signature-based and statistical anomaly detectors are implemented within a computer network to detect threats. The precision of the alerts passed to cyber-security analysts could be increased by studying the correlation structure between such detectors. Statistically, this challenge consists in estimating causal relationships between point processes of alerts. To this end, we extend a recently proposed class of models for dynamic networks called mutually exciting point process graphs, which allow for an unknown latent graph structure between node-specific point processes, where each node in the graph is a detector. Furthermore, different classes of users might be associated with different dependency structures across alerts. Motivated by these concerns, we further extend mutually exciting point process graphs to allow for the estimation of group-specific graphs of dependencies between detectors, with the goal of quantifying when individual user activity is unlikely based on the behaviour of similar users within the network.

CO045 Room Aula E NOVEL STATISTICAL METHODS FOR CENSORED AND SKEW DATA — Chair: Victor Hugo Lachos Davila

C0153: **A repairable system subjected to hierarchical competing risks: Modeling and applications**
Presenter: **Francisco Louzada**, University of Sao Paulo, Brazil

The proposal is to model for a single repairable system with a hierarchical structure under the assumption that the failures follow a nonhomogeneous Poisson process (which corresponds to minimal repair action) with a power-law intensity function. The properties of the new model are discussed in detail. The parameter estimators are obtained using the maximum likelihood method. A simulation study is conducted to show that our estimators are bias-free. The proposed modelling is illustrated on a dataset on a traction system of an in-pipe remotely controlled robotic unit.

C0216: **Moments and random number generation for the truncated elliptical family of distributions**
Presenter: **Christian Eduardo Galarza Morales**, Escuela Superior Politecnica del Litoral, Ecuador
Co-authors: Katherine Valeriano, Larissa Matos

An algorithm is proposed to generate random numbers from any member of the truncated multivariate elliptical family of distributions with a strictly decreasing density generating function. Based on previous work, we construct an efficient sampling method by means of a slice sampling algorithm with Gibbs sampler steps. We also provide a faster approach to approximate the first and the second moment for the truncated multivariate elliptical distributions where Monte Carlo integration is used for the truncated partition and explicit expressions for the non-truncated part. Examples and an application to environmental spatial data illustrate its usefulness. Methods are available for free in the new R library relliptical.

C0325: **Linear models for multivariate repeated measures data from a skew normal distribution**
Presenter: **Anuradha Roy**, The University of Texas at San Antonio, United States
Co-authors: Timothy Opheim

The theory of linear models is generalized for doubly multivariate data from matrix-variate normally distributed errors to matrix-variate skew normally distributed errors. In addition, we assume that the covariance matrix defining the location-scale matrix-variate skew normal distribution has a block compound symmetry structure. We derive the maximum likelihood estimators of the model's parameters, the Fisher information matrix for the direct, working, and centered parametrizations, and Rao's score tests and likelihood ratio tests for model building tests of hypotheses and a hypothesis test for the centered intercept. A profiling argument is used to reduce the dimensionality of the optimization method used to obtain the maximum likelihood estimators. Finally, we provide a real-world example to illustrate these derivations.

C0536: **Censored autoregressive regression modeling using the R package ARCensReg**
Presenter: **Fernanda Schumacher**, The Ohio State University, United States
Co-authors: Katherine Andreina Loor Valeriano, Victor Hugo Lachos Davila, Larissa Avila Matos, Christian Eduardo Galarza Morales

In several applications, data are collected over time and may contain censored or missing observations, making it impossible to use standard statistical procedures. The analysis of censored linear regression models with autoregressive errors is discussed using the R package ARCensReg, which implements maximum likelihood estimation via a stochastic approximation of the EM algorithm. The package was recently updated and accounts for both normal and Student-t distributions, the latter distribution being particularly relevant for dealing with data that contain outlier observations. The use of the package for model selection and estimation will be illustrated using a real data set.

CC162 Room Aula G PARAMETRIC INFERENCE — Chair: Sara Taskinen

C0270: **Parametric estimation of tempered stable laws**
Presenter: **Till Massing**, University of Duisburg-Essen, Germany

Tempered stable distributions are frequently used models in financial applications (e.g., for option pricing) in which the tails of stable distributions are too heavy. Unfortunately, given the non-explicit form of the probability density function, estimation relies on numerical algorithms such as the fast Fourier transform, which typically are time-consuming. We compare several parametric estimation methods, such as the maximum likelihood method and different generalized methods of moment approaches. We conduct extensive simulation studies to analyze finite sample properties measured by the empirical bias and precision and compare computational costs. Additionally, we study large sample properties and derive theoretical results for consistency, asymptotic normality, and asymptotic efficiency for our estimators. We cover various relevant subclasses of tempered stable distributions, including the classical tempered stable distribution and the tempered stable subordinator. Moreover, we discuss the normal tempered stable distribution, which arises by subordinating a Brownian motion with a tempered stable subordinator. Our financial application to energy spot prices illustrates the benefits of tempered stable models. The implemented routines will be published in an R package.

C0344: **Geometric goodness of fit measure to detect patterns in data point clouds**
Presenter: **Alberto Hernandez**, Universidad de Costa Rica, Costa Rica

A geometric goodness-of-fit index is derived which is similar to R^2 using geometric data analysis techniques. We build the alpha shape complex from the data cloud projected onto each variable and estimate the area of the complex and its domain. We create an index that measures the difference in area between the alpha shape and the smallest squared window of observation containing the data. By applying ideas similar to those found in the closest neighbor distribution and empty space distribution functions, we can establish when the characterizing geometric features of the point set emerge. This allows for a more precise application for our index. We provide some examples with anomalous patterns to show how our algorithm performs.

C0505: **Fast and consistent inference in compartmental models of epidemics using Poisson approximate likelihoods**
Presenter: **Michael Whitehouse**, University of Bristol, United Kingdom

Addressing the challenges of scaling up epidemiological inference to complex and heterogeneous populations, we introduce Poisson Approximate Likelihood methods for stochastic compartmental models. A Poisson Approximate Likelihood can be evaluated using only elementary linear-algebraic operations, requires no simulation from the model in order to circumvent the intractability of the true likelihood, and incurs a computational complexity which scales with the number of compartments similarly to that of forwarding Euler discretization of the corresponding ordinary differential equation model. We prove the consistency of the maximizer of the Poisson Approximate Likelihood in the regime where the population size tends to infinity. This appears to be the first consistency result concerning the large population regime for any likelihood or approximate likelihood-based estimator which is applicable across the broad class of compartmental models we consider. Through examples we demonstrate how Poisson Approximate Likelihoods can be: embedded within Delayed Acceptance Particle Markov Chain Monte Carlo to facilitate speed-ups in exact Bayesian inference; applied to an age-structured model of influenza, easily implemented in STAN and compared to ordinary differential equation models; and used to calibrate a large-scale spatial meta-population model of measles transmission.

C0626: **Penalized power-generalized Weibull distributional regression**
Presenter: **Laura McQuaid**, University of Limerick, Ireland
Co-authors: Shirin Moghaddam, Kevin Burke

Multi-parameter regression (MPR) modelling refers to the approach whereby covariates enter a parametric model through multiple distributional parameters simultaneously (e.g., scale and shape parameters), allowing more complex covariate effects to be captured. Standard techniques allow for one parameter, usually the scale, to be a function of covariates but this may result in the potential impact of the shape parameter on the hazard to be lost. Having multiple parameters depending on covariates may lead to a computationally expensive variable selection procedure. On the other hand, penalized estimation procedures such as the least absolute shrinkage and selection operator (LASSO) and adaptive LASSO are commonly used to perform variable selection and estimation simultaneously but they have primarily been developed for classical regression problems where covariates enter only through a single distributional parameter. We introduce a flexible penalized multi-parameter modelling framework and investigate its performance through simulation studies and real data analysis. We particularly focus on the three-parameter (one scale, two shapes) Power Generalized Weibull (PGW) distribution. The PGW distribution encompasses key shapes of hazard function (constant, increasing, decreasing, up then down, down then up) and a variety of common survival distributions (Weibull, log-logistic, Gompertz). This allows for a highly flexible approach for modelling survival data.

CC157 **Room Aula H** **APPLIED STATISTICS AND DATA ANALYSIS** **Chair: Qing Pan**

C0465: **Comparing dominance of tennis' big three via multiple-output Bayesian quantile regression models**
Presenter: **Bruno Santos**, University of Kent, United Kingdom

Tennis has seen a myriad of great male tennis players throughout its history and we are often interested in the discussion of who is/was the greatest player of all time. While we do not try to answer this question, we delve into comparing some key statistics related to dominance over their opponents for the male players with the most Grand Slam titles, currently: Nadal, Djokovic and Federer. We consider the minutes played and the relative points in each of their completed matches, as a measure of dominance against other players. We consider important covariates such as surface, win or loss, type of tournament and whether their opponent was a top 20 ranked player in the world or not. We consider a Bayesian quantile regression model for multiple-output response variables to take into account the dependence between minutes and relative points won. This approach is compelling since we do not need to choose a probability distribution for the joint probability distribution of our response variable. The results agree with the common intuition of Nadal's superiority on clay courts, Federer's superiority on grass courts and Djokovic's superiority on hard courts given their success on each of these surfaces; though Nadal's dominance in clay court games is unique. Federer shows his dominance regarding minutes spent on the court in wins, while Djokovic takes the edge when considering the dimension of relative points won, for most of the comparisons.

C0478: **Spatial modelling road accidents in Besancon (France) using log-gaussian cox processes**
Presenter: **Cecile Spychala**, Universite de Franche-Comte - Lmb, France
Co-authors: Camelia Goga, Clement Dombry

In order to prevent and/or forecast road accidents, the statistical modelling of spatial dependence and potential risk factors is a major asset. The focus is on the georeferenced location of accidents. We crossed these events with covariates characterizing the study geographical area such as sociodemographic and infrastructure measures. After a variable selection (Poisson model, Poisson models aggregation and random forest), the occurrence of accidents was modelled by using a spatial log-Gaussian Cox process. The results of this analysis enable us to identify principal risk factors for road accidents and critical areas. The data used are road accidents that occurred between 2017 and 2019 in the CAGB (urban community of Besancon).

C0584: **Elastic-net for instrumental variables regression**
Presenter: **Alena Skolkova**, CERGE-EI, Czech Republic

The purpose is to investigate the relative performance of the lasso, ridge and elastic-net estimators in obtaining first-stage predictions for IV estimation. Although the lasso estimator is currently established as the most popular regularization technique for prediction problems under the sparsity assumption, its performance under high correlation and grouping between instruments can be improved via the elastic-net. A Monte Carlo study demonstrates that in all analyzed scenarios the elastic-net estimator dominates other estimators in terms of the mean squared error and overall stability of the first-stage predictions. We also compare the relative performance of IV estimators that employ the lasso, ridge and elastic-net first-stage estimates. Finally, we confirm the superior performance of the elastic-net estimator in an empirical application with many IVs.

C0658: **Data science education for developing countries: The process of democratization**

Presenter: **Fulya Gokalp Yavuz**, Middle East Technical University, Turkey

Industry and academia have been attempting to develop a new science for data exploration over the last three decades. Traditional institutions indeed possess an environment conducive to the generation of new sciences and fields. However, the field of data science was itself born of developments in both industry and research institutions. Thus, it may not be realistic or correct to expect the university or department curriculum to change completely and immediately to accommodate the needs of currently evolving fields. This issue of institutional adaptation to new areas of inquiry is, moreover, more complex in developing countries than in developed ones for various reasons including a more entrenched bureaucracy and lower incomes at universities. There is thus an inevitable need for additional tools and formations to support those studying in developing countries to compete with their peers. The democratization of data science education is discussed here via a case study carried out at a major research institution in Turkey. The study demonstrates that while the field is undergoing this development, students can still adapt to this discipline with more effort in their early careers.

CC219 **Room Aula F** **FEATURE SELECTION AND VARIABLE IMPORTANCE** **Chair: Karel Hron**

C0415: **Best subset selection via continuous optimization**

Presenter: **Benoit Liquet**, Macquary University, Australia

Co-authors: Sarat Moka, Houying Zhu, Samuel Muller

Recent rapid developments in information technology have enabled the collection of high-dimensional complex data including in engineering, economics, finance, biology, and health sciences. High-dimensional means that the number of features is large and often far higher than the number of collected data samples. Several optimization and search methods have been proposed in the literature to tackle the problem of identifying or selecting the set of important predictors. These methods include forward stepwise, Lasso, and mixed-integer optimization. We will briefly review existing methods, and then present an L0 continuous optimization-based solution, a novel approach that tackles the challenging task of best subset selection for linear models, especially when the number of features is very large. Simulation results are presented to highlight the performance of the proposed method in comparison to the existing methods. Our new formulation for best subset selection in linear regression models promises to open new research avenues for feature extraction for a large variety of models.

C0506: **Domain selection for Gaussian processes**

Presenter: **Nicolas Hernandez**, UCL, United Kingdom

Co-authors: Gabriel Martos

A novel domain selection methodology is proposed for high-dimensional Gaussian processes. We use the Kullback-Leibler divergence to introduce a divergence curve as a tool to select the domain of the largest divergence between two processes. The proposed method learns and infers about the subinterval of the domain that better discriminates the classes of functions. Throughout a Monte Carlo experiment, we show the accuracy of the proposed method in the estimation of the true domain of largest divergence.

C0221: **Features selection and combination in high-dimensional data with the penalized Youden index**

Presenter: **Claudio Junior Salaroli**, University Complutense of Madrid, Spain

Co-authors: Maria del Carmen Pardo

In high-dimensional classification contexts, like with -omics data, with thousands of biomarkers and dozens of observations, it is crucial to combine regressors omitting the noise caused by thousands of irrelevant features. To achieve this task, regularization techniques are very popular methods that, adding a penalization term to the original optimization problem, allow us to achieve a sparse estimation, improving classification performances and interpretability of the result. The application of these techniques to the Youden index function, i.e. the distance between the ROC curve and the chance line, is proposed. The resulting new methodology, named Penalized Youden Index Estimator (PYE), allows to select and combine biomarkers simultaneously in a high-dimensional context, also identifying the optimal cut-off point. One additional improvement is given by considering the cut-off point as a function of specifics of the patient, like sex, age, habits such as smoking or sports activity, and so on, named covariates. This upgraded version of PYE has been called Penalized Youden index Estimator with Covariate adjusted cut-off point, or cPYE. The performances of these new approaches are compared with some popular existing methods, showing top performances in both selection and combination.

C0447: **Permutation based variable importance determination for deep learning**

Presenter: **Matthias Medl**, University of Natural Resources and Life Sciences, Vienna, Austria

Co-authors: Matthias Medl, Theresa Scharl, Astrid Duerauer, Friedrich Leisch

Statistical models with the capability to predict process variables which cannot be measured in real time have become an effective tool to monitor biopharmaceutical production processes. The implementation of novel measurement devices with the capacity to capture a wide array of physical properties of process intermediates online has led to the expansion of the variable space available to generate these models. However, extracting all information contained within this high-dimensional variable space presents a challenge. To overcome this challenge, we propose a deep-learning framework capable of processing the whole variable space to estimate critical process parameters in real time, e.g. product or impurity concentrations of a biopharmaceutical purification process. The models consist of two parallel strands that are later concatenated. One strand leverages the pattern recognition capabilities of convolutional layers to process spectral data, while the other one processes single-variable measurements and contains fully connected layers. In order to gain insight into the inner workings of the models, a permutation-based methodology has been developed to estimate the variable importance for each time point throughout the process. The variable importance workflow has subsequently been validated on artificial data with a similar data structure, where the variable importance has been predefined and was thus known.

Tuesday 23.08.2022 16:15 - 17:45 Parallel Session D – COMPSTAT2022

CV193 Room Aula B APPLIED STATISTICS (VIRTUAL) Chair: Anuradha Roy

C0502: **A statistical approach to evaluate last minute pricing decisions in the online hotel market**
Presenter: **Andrea Guizzardi**, Alma Mater Studiorum University of Bologna, Italy
Co-authors: Luca Vincenzo Ballestra, Enzo DInnocenzo

A nonlinear statistical framework is proposed for studying the last-minute pricing decisions of hotels active in the online market. In particular, the last-minute rate is specified as a linear function of the early booking rate and a shock term. The latter is regarded as the combined effect of the hoteliers forecasting error about the pick-up curve and their last-minute pricing tactics. We connect the parameters of the shock distribution to the revenue managers' practices, modeling location, scale, skewness and kurtosis with a dynamic score-driven approach. To overcome possible issues of endogeneity, a nonlinear instrumental variable estimator is employed. An empirical analysis is performed where we leverage a large dataset obtained by scraping information that is publicly available on the internet. Results show that the error term is accurately described by a skew-t distribution, rather than by a Gaussian specification. Moreover, the score-driven model turns out to be very suitable for capturing the complex nonlinear behavior of online everyday pricing decisions. Actually, the proposed approach is a reliable and transparent tool to assess the online pricing behavior of any hotel that publishes rates on the internet.

C0592: **Estimation and inference on the partial volume under the ROC surface**
Presenter: **Katherine Young**, University of Kansas Medical Center, United States
Co-authors: Leonidas Bantis

Summary measures of biomarker accuracy that employ the receiver operating characteristic (ROC) surface have been proposed for biomarkers that classify patients into one of three groups: healthy, early-stage, or advanced-stage disease. The well-known volume under the ROC surface (VUS) summarizes the overall discriminatory ability of a biomarker in such configurations. However, the VUS includes thresholds associated with clinically irrelevant true classification rates (TCRs). For example, due to the lethal nature of pancreatic cancer, thresholds associated with a low TCR for identifying patients with pancreatic cancer may be undesirable and not appropriate for use in a clinical setting. We study the properties of a more focused criterion, the partial volume under the ROC surface (pVUS), that summarizes the diagnostic accuracy of a marker in the three-class setting for regions restricted to only those of clinical interest. We propose methods for estimation and inference of the pVUS under parametric and non-parametric frameworks and apply these methods to the evaluation of potential biomarkers for the diagnosis of pancreatic cancer.

C0636: **Failure rate monitoring in generalized gamma-distributed process**
Presenter: **Niladri Chakraborty**, University of the Free State, South Africa
Co-authors: Tahir Mahmood

Technological advancement has brought revolutionary changes in product quality in today's time. Most of the manufacturing processes produce a large number of conforming items along with a few nonconforming items. In real-time monitoring of these highly efficient processes, monitoring time between events is a well-known approach in statistical process control literature. It is generally assumed that the time-between-events follows an exponential or gamma distribution. However, the generalized gamma distribution is often a more popular choice in modelling skewed data. In this context, we consider a two-sided monitoring scheme based on the generalized gamma distribution. Two-sided monitoring schemes for skewed distributions often encounter bias in their run length properties. Therefore, we address this problem with modified control limits in a more general distributional setup. A Monte-Carlo simulation-based study is designed, and computational results reveal encouraging performance properties. Practical applications related to monitoring renewable energy production and coal mine explosions have been considered.

C0614: **A novel environmental system-focused empirical mode decomposition analysis: Application to Minas passage**
Presenter: **Ian Kenny**, The Open University, United Kingdom
Co-authors: Dhouha Kbaier

The application of Empirical Mode Decomposition (EMD) to four components of the Minas Passage data is considered. Specifically, we present a finding which places the number of Intrinsic Mode Functions (IMFs) as a function of the number of inputs, rather than the number of observations N. The previous empirical finding that the number of IMF required was not more than $\log 2N$ meant that the number of functions required grew with the number of observations. In addition, being able to relate the number of functions required to the number of inputs has the effect of enabling the identification of the systemic relationships within the signal, this is achieved through the judicious application of Time-Dependent Intrinsic Correlation (TDIC) to the identified IMFs. The discovery of an input-related empirical measure for the number of IMFs also addresses the issue of mode mixing, by viewing the number of modes in relation to the system rather than the number of observations. Therefore, the smaller signal components are treated as if they are artefacts of the interactions between the signals. By applying this method, we have been able to show that EMD can be used as a starting point for building system-focused adaptive models

CI015 Room Aula F BAYESIAN AND COMPUTATIONAL EXTREME VALUE ANALYSIS Chair: Miguel de Carvalho

C0287: **A Bayesian non-parametric approach for multivariate peak over threshold models and anomaly detection**
Presenter: **Bruno Sanso**, University of California Santa Cruz, United States
Co-authors: Peter Trubey

A constructive definition of the multivariate Pareto is considered that factorizes the random vector into a radial component and an independent angular component, using the infinity norm. We propose a method for inferring the distribution of the angular component whose support is the limit of the positive orthants of the unit p-norm spheres. We introduce a projected gamma family of distributions defined as the projection of a vector of independent gamma random variables onto the p-norm sphere. This family serves as a building block for a flexible family of distributions obtained as a Dirichlet process mixture of projected gammas. For model assessment and comparison, we discuss model scoring methods appropriate to distributions on the unit hypercube. In particular, working with the energy score criterion, we develop a kernel metric appropriate to the infinity norm unit hypercube that produces a proper scoring rule. We leverage this score for the detection of observations that are anomalous when compared to their predictive distribution. We consider simulated data, as well as data corresponding to integrated vapor transport (IVT), a variable that describes the rate of flow of moisture in the atmosphere along the coast of California for the years 1979 through 2020. We find a clear but heterogeneous geographical dependence.

C0383: **An extreme value Bayesian Lasso for the conditional bulk and tail**
Presenter: **Miguel de Carvalho**, FCiencias.ID - Associacao para a Investigacao e Desenvolvimento de Ciencias, Portugal
Co-authors: Patricia de Zea Bermudez

A novel regression model for the conditional bulk and conditional tail of a possibly heavy-tailed response is introduced. The proposed model can be used to learn the effect of covariates on an extreme value setting via a Lasso-type specification based on a Lagrangian restriction. The model can be used to track if some covariates are significant for the bulk, but not for the tail—and vice-versa; in addition to this, the proposed model avoids the need for conditional threshold selection in an extreme value theory framework. The finite-sample performance of the proposed methods is assessed by means of a simulation study that shows that our method recovers the true conditional distribution over a variety of simulation scenarios, along with being accurate in variable selection. Rainfall data are used to display how the proposed method can learn to distinguish between key drivers of moderate rainfall, against those of extreme rainfall.

C0386: **Distributed inference for extreme value analysis of large spatial datasets**
Presenter: **Emily Hector**, North Carolina State University, United States

Extreme environmental events frequently exhibit spatial and temporal dependence. These data are often modeled using max stable processes that are computationally prohibitive to fit for as few as a dozen observations. We propose a spatial partitioning approach based on local modeling of subsets of the spatial domain that delivers computationally and statistically efficient inference. The proposed distributed approach is extended to estimate spatially varying coefficient models to deliver computationally efficient modeling of spatial variation in marginal parameters. We illustrate the flexibility of our approach through simulations and the analysis of streamflow data from the U.S. Geological Survey.

CO131 **Room Aula G** ANALYSIS OF COMPLEX REAL LIFE DATA **Chair: Qing Pan**

C0412: **Risk predictions using panel count data with informative observation times**
Presenter: **Qing Pan**, George Washington University, United States

In epidemiology studies of screening-detected disease, researchers often face screening data in the form of interval censored panel count data. Furthermore, observation times are usually informative about the disease risks. We analyze the Study of Colonoscopy Utilization within the PLCO Cancer Screening Trial, which followed patients for up to 15 years on colorectal cancer screening results including both cancer and its non-advanced/advanced adenoma precursors. Screening times strongly depend on past screening results. Recurrent adenoma processes are defined by the numbers and sizes of adenoma. Furthermore, the recurrent adenoma processes are reset to zero at each screening because colonoscopy removes all detected adenomas. We model the recurrent times to screening and recurrent adenoma at each screening jointly. Correlations between the screening and adenoma processes are modeled by subject-specific frailty terms. The baseline intensity function and regression coefficients for the recurrent adenoma processes are estimated using estimating equations for interval censored panel count data under the piecewise baseline intensity assumption. Probabilities of advanced adenoma at the next fixed or expected screening time are predicted. Performance of the risk prediction is examined through extensive simulation studies and illustrated on the PLCO clinical trial data.

C0435: **Informed presence in electronic health record data: Bias and bias reduction approaches in longitudinal analyses**
Presenter: **Yun Li**, University of Pennsylvania, United States

Electronic health record (EHR) systems capture patient information inconsistently, with patients contributing more data when they are sick than when they are healthy. This creates "informed presence" and systematic differences between captured and non-captured data biasing estimates of association. There is growing interest in analytic approaches that account for informed presence, but practical approaches for conceptualizing, identifying, and accounting for informed presence in applied EHR-based research have received limited attention. We introduce a collider-bias framework for understanding informed presence bias, novel visualization strategies for irregularly measured data, and four approaches to bias reduction under informed presence. To illustrate, we investigated associations between steroids and cytomegalovirus viremia among pediatric solid organ transplant patients ($N = 271$) in a recurrent outcomes analysis. We identified conceptual, descriptive, and analytic evidence of informed presence. Incidence rate ratios dropped from 1.83 (95% CI: 1.02, 3.28) in a naive analysis to 1.37 (0.73, 2.57) when accounting for informed presence using inverse intensity weighting. When conducting analyses with irregularly measured EHR data, we recommend: 1) identifying the expected observation process using conceptual diagrams; 2) visualizing dependence in the observation process; 3) and accounting for outcome dependence in analyses.

C0421: **Multiple imputation methods for functional data with applications in mental health research**
Presenter: **Adam Ciarleglio**, George Washington University, United States

In mental health research, the number of studies that include multimodal neuroimaging is growing. Often, the goal is to integrate both the clinical and imaging data to address a specific research question. Functional data analytic tools for analyzing such data can perform well, but these methods assume complete data. In practice, some proportion of the data may be missing. We present several approaches for imputation of missing scalar and functional data when the goal is to fit functional regression models for the purpose of estimating the association between a scalar or functional outcome and scalar and/or functional predictors. We present results from a simulation study showing the performance of various imputation approaches with respect to fidelity to the observed data and estimation of the parameters of interest. The proposed approaches are illustrated using data from a placebo-controlled clinical trial assessing the effect of SSRI on subjects with major depressive disorder.

C0190: **Functional modeling of telecommunications data**
Presenter: **Algimantas Birbilas**, Vilnius University, Lithuania
Co-authors: Alfredas Rackauskas

Statistical modeling and forecasting of telecommunications data are considered. Main mobile traffic events (SMS, Voice calls, Mobile data) are smoothed using B-spline functions and later analyzed in a functional framework. Functional linear auto-regression models are fitted using both bottom-up and top-down design methodologies. The advantages and disadvantages of both approaches for the prediction of mobile telephone users' habits are discussed.

CO031 **Room Aula C** STATISTICAL TEXT MINING **Chair: Peter Winker**

C0335: **Measuring fiscal policy preferences based on the German Bundestag speeches and public discourse**
Presenter: **Viktoriia Naboka**, Justus Liebig Unversity of Giessen, Germany
Co-authors: Peter Winker, Peter Tillmann, Albina Latifi

Fiscal policy, i.e. changes in government spending and tax revenues, is an important determinant of business cycles. Research efforts, so far, heavily rely on human judgment in order to measure fiscal policy shocks. Therefore, reproducible quantitative computational text analysis to quantify fiscal policy preferences are applied. The analysis is based on the driver of fiscal policy, i.e. the fiscal policy debates in the Bundestag, which are also accompanied by public discourse. Thus, two novel data sets on parliamentary speeches and newspaper reporting covering the time period from 1960 to 2021 are used. First, advanced NLP (Natural Language Processing) techniques, such as topic modelling, are applied to uncover latent topics in the corpus and to study the evolution of topic importance over time. Second, an embedding-based approach, which allows the representation of words and documents in a shared vector space, is proposed to measure fiscal policy-related sentiment. For this reason, a dictionary containing terms related to expansive and restrictive policy measures is created. Finally, an index is proposed that captures the sentiment from speeches and news articles at a scale from restrictive to expansive. In future research, the novel indicator will be used in econometric models to examine the macroeconomic impact of fiscal policy.

C0515: **Comparative analysis of LDA model selection criteria based on Monte Carlo simulations**
Presenter: **Victor Bystrov**, University of Lodz, Poland
Co-authors: Viktoriia Naboka, Anna Staszewska-Bystrova, Peter Winker

The performance of the recently developed singular Bayesian information criterion (sBIC) in selecting the number of topics in LDA models is evaluated and compared to the performance of alternative model selection criteria proposed for topic models. The sBIC is a generalization of the standard BIC that can be implemented in singular statistical models. The comparison is based on Monte Carlo simulations and carried out for several alternative settings, varying with respect to the number of topics, the number of documents and the size of documents in the corpora. Practical recommendations for LDA model selection are developed.

C0342: **Identification of innovation diffusion trends with FDA clustering**
Presenter: **Albina Latifi**, Justus Liebig University Giessen, Germany
Co-authors: Peter Winker, David Lenz

The Diffusion of Innovation Theory often describes innovation diffusion as a hump-shaped curve. Light is shed on this theory by using a data-driven approach based on news articles from a technology-related newspaper for the period 1996 - 2021. In a first step, computational methods from natural language processing such as topic modelling were used to identify latent topics in the text corpus and to obtain associated time series of topic weights. In a second step, methods from the field of functional data analysis (FDA) were applied to categorize these time series in clusters. For this purpose, the k-means method, which is often used in the literature for related tasks was compared with an implementation of the global search heuristic Threshold Accepting (TA) for clustering. Preliminary results indicate that TA provides better and more robust results than k-means. Different clusters of innovation diffusion trends were identified, suggesting that empirically there are various shapes of diffusion which do not all resemble the classical diffusion curve to the same extent. Moreover, this approach could uncover different stages of innovation diffusion. Based on these results, success predictions for individual innovations might be derived.

C0570: **A survey of scientists opinions on climate mitigation policy**
Presenter: **Ivan Savin**, Universitat Autonoma de Barcelona, Spain

There are numerous instruments available in the climate-policy toolbox, but it is contested which are the most important ones to reach the Paris climate goals. Many economists but also other researchers argue that carbon pricing should be the central climate policy instrument within climate policy mixes. While lots of studies investigated its role in the economy, both theoretically and empirically, less is known about what climate researchers from other social sciences think about this policy. To advance the debate on climate policy and particularly on carbon pricing, we present the findings of a new, global survey of scientists' views on climate policies. The aim is threefold. First, we quantify the degree of (dis)agreement over carbon pricing and several other policies across various research fields. Second, we elicit the sources of disagreement. To achieve this, we analyze how the importance of instruments is associated with weights and perceptions of distinct policy criteria. Third, we derive an interdisciplinary agenda for future research on climate policy by analyzing responses to open questions using a topic modelling method. The particular contribution of the present expert survey is to provide a detailed analysis of consensus about climate policy across a diversity of disciplines.

CO123 Room Aula D STATISTICAL ANALYSIS OF NETWORKS — Chair: Francesco Sanna Passino

C0345: **Changepoint inference with a graphical dependence structure**
Presenter: **Nick Heard**, Imperial College London, United Kingdom

When analysing multiple time series that may be subject to changepoints, it is sometimes possible to specify a priori, by means of a graph, which pairs of time series are likely to be impacted by simultaneous changes. An informative prior distribution for changepoints is introduced which encodes the information from the graph, providing a changepoint model for multiple time series that borrows strength across clusters of connected time series to detect weak signals for synchronous changepoints. The approach is extended to allow dependence between nearby but not necessarily synchronous changepoints across neighbouring time series in the graph. For inference, we use an adaptation of the partial decoupling method for auxiliary variable reversible jump MCMC. The merit of the proposed approach is demonstrated through a changepoint analysis of computer network authentication logs from Los Alamos National Laboratory (LANL), demonstrating an improvement at detecting weak signals for network intrusions across users linked by network connectivity, whilst limiting the number of false alerts.

C0358: **Spectral embedding and the latent geometry of multipartite networks**
Presenter: **Alexander Modell**, University of Bristol, United Kingdom

Graph embedding is the task of representing the nodes of a network as points in space, the geometric features of which can reveal the underlying structure of the network. We focus on spectral embedding, where this point cloud is constructed using a spectral decomposition of the graph. Typically, statistical analyses of such embeddings concern unipartite graphs, however, many real-world graphs are multipartite. Examples include networks linking users with items (a bipartite graph), drugs with diseases with genes and computers with users with processes (tripartite graphs). We demonstrate that in this case, the embedding lies close to low-dimensional subspaces of a higher-dimensional space, each corresponding to a type of node, and propose an algorithmic step which exploits this for further dimension reduction. Results are illustrated with various applications, including anti-corruption and biomedicine.

C0385: **Community detection in multilayer degree-corrected stochastic blockmodels**
Presenter: **Joshua Agterberg**, Johns Hopkins University, United States
Co-authors: Jesus Arroyo, Zachary Lubberts

In multilayer network analysis, networks often share some underlying structure (such as communities) but have possibly heterogeneous network-specific idiosyncrasies that can make simple averaging procedures fail even at the population level. We propose the multilayer degree-corrected stochastic blockmodel, a multilayer network model that assumes that communities are shared across networks, but that edge probabilities and degree corrections can vary between networks. First, we discuss why averaging procedures may fail, and then we discuss the underlying shared spectral geometry. Inspired by this geometry, we propose an algorithm to take advantage of the shared structure amongst the networks, and we show its performance on real and simulated data.

C0666: **Manifold structure in graph embeddings**
Presenter: **Patrick Rubin-delanchy**, University of Bristol, United Kingdom

Statistical analysis of a graph often starts with embedding, the process of representing its nodes as points in space. How to choose the embedding dimension is a nuanced decision in practice, but in theory, a notion of true dimension is often available. In spectral embedding, this dimension may be very high. However, it is shown that existing random graph models, including graphon and other latent position models, predict the data should live near a much lower dimensional manifold. One may therefore circumvent the curse of dimensionality by employing methods which exploit hidden manifold structures. We illustrate results on toy as well as cyber-security network data.

CO103 Room Aula I STATISTICAL METHODS FOR SURVIVAL DATA — Chair: Marialuisa Restaino

C0417: **Bias induced by ignoring double truncation**
Presenter: **Carla Moreira**, University of Minho, Portugal

Truncation is a well-known phenomenon in some observational studies of time-to-event data. For example, when the sample restricts to those individuals with events falling between two particular dates, they are subject to selection bias due to the simultaneous presence of left and right truncation, also known as interval sampling, leading to a double truncation. When time-to-event data is doubly truncated, the sampling information includes the variable of interest X and left-truncation and right-truncation variables U and V, but the observable population reduces to those individuals for which the variable of interest lies between left-truncation and right-truncation variables. In this case, both large and small values of X are observed in principle with a relatively small probability. The observational bias for X varies from application to application, depending on the joint distribution of (X,U,V). For a particular x, the probability of sampling a value x may be roughly constant, inducing no observational bias; or it may be not constant, indicating bias induced by double truncation. We present the problem of estimating the distribution of X and other related curves, using nonparametric and semiparametric approaches, from a set of iid triplets with the distribution of (X,U,V) given the double truncation condition. We present several scenarios where double truncation appears in practice and analyse the effect of ignoring double truncation in such cases.

C0522: **A note on nonparametric survival functions under censored and truncated data**

Presenter: **Marialuisa Restaino**, University of Salerno, Italy

Co-authors: Sara Milito

Survival data (univariate and bivariate) have received considerable attention recently. In survival analysis, it is common to deal with incomplete information of the data, due to random censoring and random truncation. Most of the existing research on bivariate survival analysis focuses on considering the case when components are either censored or truncation or when one component is censored and truncated, but the other one is fully observed. Starting from this background, we will review the most used estimators for the survival function (univariate and bivariate), by taking into the incomplete information due to censoring and truncation. We will study the differences between them and we will compare their performance through a simulation study and application to real datasets.

C0556: **Relative treatment effects in two dependent samples: An alternative to logrank or sign tests**

Presenter: **Dennis Dobler**, Vrije Universiteit Amsterdam, Netherlands

Relative treatment effects quantify the probability that someone who received Treatment A survives longer than someone who received Treatment B. We call Treatment A superior if that probability exceeds 50%. The underlying survival data are assumed to be independently right-censored. We will discuss different possible definitions of relative treatment effects in paired data, depending on which estimation is based on Kaplan-Meier or Aalen-Johansen estimators. Two big advantages over existing tests are that inference procedures based on relative treatment effects are still useful in the case of crossing hazards and they are based on statistics that allow for easy interpretations. A randomization method and the bootstrap are used to produce reliable inference procedures. We will apply the methods to datasets in the context of diabetes (time to blindness).

C0695: **General proportional mean residual and past lifetime frailty models**

Presenter: **Fatemeh Hooti**, University of Naples Federico II, Italy

Co-authors: Jafar Ahmadi, Maria Longobardi

The mean residual lifetime and mean past lifetime functions for a lifetime random variable are important characteristics of the model in reliability theory and survival studies. Based on these two measures, proportional mean residual and mean past lifetime models were introduced as regression models. In this work, by considering a frailty variable, the proportional mean residual life frailty model and also, general mean past lifetime frailty models are introduced. Unconditional survival and density functions of lifetime variables based on the proposed frailty model are obtained. Some stochastic ordering properties are studied.

CO176 Room Aula Q DIMENSION REDUCTION IN RECENT CROSS SECTIONAL AND TIME SERIES METHODS	**Chair: Matteo Farne**

C0434: **Testing high-dimensional general linear hypotheses under a multivariate regression model with spiked noise covariance**

Presenter: **Alexander Aue**, UC Davis, United States

Co-authors: Haoran Li, Debashis Paul, Jie Peng

The problem of testing linear hypotheses under a high-dimensional multivariate regression model with spiked noise covariance is considered. The proposed family of tests consists of test statistics based on a weighted sum of projections of the data onto the factor directions, with the weights acting as the regularization parameters. We establish the asymptotic normality of the proposed family of test statistics under the null hypothesis. We also establish the power characteristics of the tests under a family of probabilistic local alternatives and derive the minimax choice of the regularization parameters. The performance of the proposed tests is evaluated in comparison with several competing tests. Finally, the proposed tests are applied to the Human Connectome Project data to test for the presence of an association between volumetric measurements of the human brain and certain behavioral variables.

C0517: **The maximum of the periodogram of a sequence of functional data**

Presenter: **Vaidotas Characiejus**, University of Southern Denmark, Denmark

Co-authors: Clement Cerovecki, Siegfried Hoermann

The detection of periodic signals in functional time series is investigated when the length of the period is not assumed to be known. A natural test statistic for the detection of periodicities is the maximum overall fundamental frequencies of the Hilbert-Schmidt norm of the periodogram operator. Using recent advances in Gaussian approximation theory, we show that under certain assumptions the appropriately standardised test statistic belongs to the domain of attraction of the Gumbel distribution. The asymptotic results allow us to construct tests for hidden periodicities. We demonstrate the performance of our methodology in a simulation study and we also illustrate the usefulness of our approach by examining periodicities in the air quality data from Graz, Austria and showing that our approach is not only able to detect the presence of periodic signals but it is also able to reveal the structure of periodicities in the data.

C0670: **An algebraic estimator for large spectral density matrices**

Presenter: **Matteo Farne**, University of Bologna, Italy

Co-authors: Matteo Barigozzi

A new estimator of high-dimensional spectral density matrices is introduced which is called UNshrunk ALgebraic Spectral Estimator (UNALSE), under the assumption of an underlying low rank plus sparse structure, as typically assumed in dynamic factor models. The UNALSE is computed by minimizing a quadratic loss under a nuclear norm plus l_1 norm constraint to control the latent rank and the residual sparsity pattern. The loss function requires as input the classical smoothed periodogram estimator and two threshold parameters, the choice of which is thoroughly discussed. We prove the consistency of UNALSE as both the dimension p and the sample size T diverge to infinity, as well as algebraic consistency, i.e., the recovery of latent rank and residual sparsity pattern with probability one. The finite sample properties of UNALSE are studied by means of an extended simulation exercise as well as an empirical analysis of US macroeconomic data.

C0525: **Pattern recovery by SLOPE**

Presenter: **Malgorzata Bogdan**, Lund University, Sweden

Co-authors: Xavier Dupuis, Piotr Graczyk, Bartosz Kolodziejek, Tomasz Skalski, Patrick Tardivel, Maciej Wilczynski

LASSO and SLOPE are two popular methods for dimensionality reduction in high-dimensional regression. LASSO can eliminate redundant predictors by setting the corresponding regression coefficients to zero, while SLOPE can additionally identify clusters of variables with the same absolute values of regression coefficients. It is well known that LASSO Irrepresentability Condition is sufficient and necessary for the proper estimation of the sign of sufficiently large regression coefficients. We formulate an analogous Irrepresentability Condition for SLOPE, which is sufficient and necessary for the proper identification of the SLOPE pattern, i.e. of the proper sign as well as of the proper ranking of the absolute values of individual regression coefficients, while proper ranking guarantees a proper clustering. We also provide asymptotic results on the strong consistency of pattern recovery by SLOPE when the number of columns in the random design matrix is fixed while the sample size diverges to infinity.

CO095 Room Aula E STATISTICAL LEARNING IN PRACTICE **Chair: Alejandro Murua**

C0590: **Reduced-rank tensor-on-tensor regression and tensor-variate analysis of variance**
Presenter: **Ranjan Maitra**, Iowa State University, United States
Co-authors: Carlos Llosa-Vite

Fitting regression models with many multivariate responses and covariates can be challenging, but such responses and covariates sometimes have a tensor-variate structure. We extend the classical multivariate regression model to exploit such a structure in two ways: first, we impose four types of low-rank tensor formats on the regression coefficients. Second, we model the errors using the tensor-variate normal distribution that imposes a Kronecker separable format on the covariance matrix. We obtain maximum likelihood estimators via block-relaxation algorithms and derive their computational complexity and asymptotic distributions. Our regression framework enables us to formulate a tensor-variate analysis of variance (TANOVA) methodology. This methodology, when applied in a one-way TANOVA layout, enables us to identify cerebral regions significantly associated with the interaction of suicide attempters or non-attemptor ideators and positive-, negative- or death-connoting words in a functional Magnetic Resonance Imaging study. Another application uses three-way TANOVA on the Labeled Faces in the Wild image dataset to distinguish facial characteristics related to ethnic origin, age group and gender.

C0591: **Unsupervised deep learning of ATAC-seq peaks**
Presenter: **Karin Dorman**, Iowa State University, United States
Co-authors: Yudi Zhang, Ha Thi Hong Vu, Geetu Tuteja

ATAC-seq (Assay of Transposase Accessible Chromatin sequencing) is widely used to identify open regions in the genome by "calling peaks" where sequenced DNA fragments, accessed and cut by a transposase, are enriched. Most unsupervised peak calling methods are based on traditional statistical models and suffer from elevated false positive rates. Newly developed supervised deep learning methods can be successful, but they rely on high-quality labeled data, which can be difficult to obtain. Neither approach considers biological replicates. We propose a novel deep learning method that uses unsupervised contrastive learning to extract shared signals from two or more replicates. Raw coverage data are encoded to obtain low-dimensional embeddings, optimized to minimize a contrastive loss over biological replicates. In addition, the embeddings produce peak predictions, again under a contrastive loss, and decoded to denoised data under an autoencoder loss. We compare our method with the unsupervised methods MACS2 and HMMRATAC on human ATAC-seq data, using labels obtained from related ChIP-seq experiments as a noisy truth. Our method is more precise (fewer false positives) than competing unsupervised methods. It is also a more effective denoiser of low-quality ATAC-seq data than ATACWorks.

C0501: **Evaluating probabilistic classifiers: The Triptych**
Presenter: **Timo Dimitriadis**, Heidelberg University, Germany
Co-authors: Tilmann Gneiting, Alexander Jordan, Peter Vogel

Predicting the occurrence probability of binary events is presumably the most common forecasting task throughout the sciences. Hence, a unified methodology for evaluating and comparing these forecasts is of great importance. We propose a new "triptych" of evaluating displays consisting of the receiver operating characteristic (ROC) curve, the CORP reliability diagram, and the Murphy diagram. Individually, these three displays focus on different and complementary aspects of the forecast's performance. The ROC curve assesses discrimination, the reliability diagram evaluates calibration, and the Murphy diagram combines both properties and visualizes overall predictive ability. In combination, these displays visualize the full generality of a forecast's predictive ability. This intuition is supported by showing the first theoretical result connecting these plots in full generality: For auto-calibrated forecasts, the ROC curve and the Murphy diagram display congruent information. We illustrate our proposal through four case studies ranging from astrophysics, meteorology, economics, and social science.

C0599: **Independent Metropolis sampler without rejection**
Presenter: **Florian Maire**, universite de montreal, Canada

In Bayesian statistics, approximating the posterior distribution of the model parameters is key to multiple methods of estimating posterior expectations. We consider the situation where (i) simulating i.i.d. samples from the approximate posterior is doable but computationally expensive and (ii) evaluating the unnormalized posterior density is also computationally expensive. We show that there exists a modification of the Independent Metropolis sampler which is particularly useful in such contexts: central to this construction is the idea that proposed candidates cannot be rejected but rather always accepted at the "right" time in the dynamic of the Markov chain. We will illustrate the benefit of such an algorithm for the problem of identifying therapeutic strategies that bolster antitumor immunity. In this model, the likelihood evaluation involves solving a system of ordinary differential equations that has to be numerically integrated.

CC160 Room Aula H MACHINE LEARNING AND DATA SCIENCE **Chair: Elisa Perrone**

C0252: **Compression-enabled interpretability of deep learning models for scientific discovery**
Presenter: **Reza Abbasi Asl**, University of California, San Francisco, United States

In the past decade, research in machine learning has been exceedingly focused on the development of models with remarkably high predictive capabilities. Specifically, models based on deep learning principles have shown promise in scientific discovery within domains such as neuroscience and healthcare. However, the huge number of parameters in these models have made them difficult to interpret for domain experts. We will discuss the role of model compression in building more interpretable and more stable deep learning models in the context of two computational neuroscience studies. First, we will introduce a group of iterative model compression algorithms for deep learning models. We will then discuss their role in building interpretable voxel-wise models of human brain activity evoked by natural movies. These compressed models reveal increased category-selectivity along the ventral visual pathway in human visual cortex with higher stability compared to uncompressed models. Then, we will investigate a compression-enabled stability-driven model interpretation framework to characterize complex biological neurons in non-human primate visual cortex. This visualization uncovers the diversity of stable patterns explained by neurons. Overall, these findings suggest the importance of model compression for stability-driven interpretation of deep learning models in scientific applications.

C0533: **Kernelised Stein discrepancy for truncated density estimation**
Presenter: **Daniel Williams**, University of Bristol, United Kingdom
Co-authors: Song Liu

Often, observations are truncated by a pre-defined boundary. For example, if we want to analyse the storm location pattern in the USA, the observations are naturally limited to the area within the country's boundary. Estimating a truncated density model is difficult due to the intractable normalising constant: It ensures the density model integrates to one over the truncated domain. A score matching-based approach has been proposed to estimate truncated density models using a weighted objective function. However, it is unclear whether its weighting function is optimal for a specific dataset. A truncated kernelised Stein discrepancy (TKSD) approach is developed, which solves the density estimation problem in an entirely data-driven fashion. It adapts the regular kernelised Stein discrepancy (KSD) estimator to the truncated observation setting without a handpicked weighting function. By solving the Lagrangian dual optimisation problem, we develop an objective function that estimates any unnormalised density, including one with a truncated domain. Finally, experiments on toy and real-world datasets show the accuracy of our method achieves a convincing lead over previous works at a small cost of computation time.

C0564: **Estimating continuous-time Markov chain transition rate functions with neural networks**

Presenter: **Majerle Reeves**, University of California, Merced, United States

Co-authors: Harish Bhat

Continuous-time Markov chains are used to model stochastic systems where transitions can occur at irregular times, e.g., chemical reaction networks, population dynamics, and gene regulatory networks. We develop a method to learn a continuous-time Markov chain's transition rate functions from a fully observed time series. In contrast with existing methods, our method allows for transition rates to depend nonlinearly on both state variables and external covariates. The Gillespie algorithm is used for generating trajectories of stochastic systems where propensity functions (reaction rates) are known. Our method can be viewed as the inverse: given trajectories of a stochastic reaction network, we generate estimates of the propensity functions. While previous methods used linear or log-linear methods to link transition rates to covariates, we use neural networks, increasing the capacity and potential accuracy of learned models. In the chemical context, this enables the method to learn propensity functions from non-mass-action kinetics. We test our method with synthetic data generated from a variety of systems with known transition rates. We show that our method learns these transition rates accurately, both in terms of mean absolute error between ground truth and learned transition rates, and in terms of the statistics of true and predicted trajectories.

C0578: **Generative neural networks via scoring rule minimization for probabilistic forecasting and likelihood-free inference**

Presenter: **Lorenzo Pacchiardi**, University of Oxford, United Kingdom

Co-authors: Ritabrata Dutta

Generative neural networks represent probability distributions by transforming samples from a simple base measure via a flexible transformation parametrized by a neural network. Unfortunately, they do not allow evaluating the probability density but only sampling from it, which makes training by maximum likelihood unfeasible. Usually, therefore, such neural networks are fit to a set of samples using adversarial training, which involves iteratively optimizing a min-max objective. This procedure is unstable and often leads to a learned distribution underestimating the uncertainty - in extreme cases collapsing to a single point. We discuss training generative networks via scoring rule minimization, an overlooked adversarial-free method which allows smooth training and leads to better uncertainty quantification. We show applications of this method to probabilistic forecasting and Bayesian likelihood-free inference; in both cases, the scoring rule approach leads to better performances in shorter training time.

Tuesday 23.08.2022 17:55 - 18:55 Parallel Session E – COMPSTAT2022

CO085 Room Aula D APPLIED DATA SCIENCE AND STATISTICAL LEARNING Chair: Frederic Bertrand

C0314: **Reinforcement learning for next best action recommendation in process data**
Presenter: **Yoann Valero**, Universite de Technologie de Troyes, France
Co-authors: Leonard Arnold Ebongue Ebaha, Frederic Bertrand, Myriam Maumy

Predictive business process monitoring (PBPM) aims at predicting the future of running process instances, be it for the next event or the remaining sequence of events (suffix). An event is characterized at least by a triplet made of a unit identifier, an activity the unit can go through, and the time of activity execution. The aim is to set some stepping stones for reinforcement learning in PBPM. Indeed, When making predictions for processes, there is no guarantee that these predictions will be advantageous. Thus, we have developed a method allowing for the next best action recommendation using an agent-based reinforcement learning method. This method has proven to be viable, even with low amounts of data points. In addition, this algorithm allows the recommendation of the entire remaining activities and avoids sub-optimal paths for ongoing units. We tested this algorithm on both public and real-life business process data successfully.

C0320: **Sensitivity analysis using discrete event simulation on the selling times of a fraction of a stock**
Presenter: **Elizaveta Logosha**, Universite de technologie de Troyes, Vivialys Groupe, France
Co-authors: Frederic Bertrand, Myriam Maumy

Process simulation is a technique used in various branches of industry. When speaking about the commercial process, less attention is paid to machine capacity or resource utilization, and more to the generation of customers, their transit times between the process steps such as discovery, proposal, purchase, and the prediction of the required number of sales. The construction of a theoretic model allows representing these sales and the state of the stock in real time through the simulation of discrete events. It is necessary to determine the impact of different simulation parameters and process characteristics on the quantities required, such as the percentage of depleting stock. An example of a stock is the number of flats in a future building. The sales process, studied by means of a process mining technique, is simulated according to these parameters in order to obtain as a result a number of different scenarios and to determine the confidence interval for the sale date of the specified percentage of the product. The study of the accuracy of the date found is completed by sensitivity analysis and determination of Sobol indices.

C0552: **Forecasting electricity consumption at household level**
Presenter: **Fatima Fahs**, University of Strasbourg, France
Co-authors: Frederic Bertrand, Myriam Maumy-Bertrand

National-scale electrical load forecasting has been a topic of research for decades. A variety of models and techniques have been successfully developed to accomplish this task. The recent massive deployment of smart meters at the household scale has stimulated research on electrical load forecasting at this scale, both in academia and in industry. An accurate forecast of load at the scale of an individual household is beneficial to both electricity providers and consumers. Electricity providers rely heavily on household-scale load forecasts to improve the efficiency of smart grid applications, such as smart home energy management systems and demand response applications. Thus, by providing their customers with advanced services, they can optimize their electricity consumption and reduce their electricity bills. Due to the high volatility of load curves at the household level, forecasting load at this scale remains a challenge for researchers, especially since the research published on this subject so far does not meet the real-world challenge. In our study, we propose daily forecast models of half-hourly electrical loads at the household level. The performance of the models is evaluated on disparate real load curves of the residential and tertiary sectors provided by a French electricity company. Also, we provide a forecasting approach for the most volatile load curves. Our study takes into account the industrial constraints to propose an industrially viable approach.

CO164 Room Aula H BIOMEDICAL RESEARCH ON BIOMARKERS: METHODS & APPLICATIONS (VIRTUAL) Chair: Laura Antolini

C0330: **Personalized response to treatment in patients with MS: Do different patients show benefits on different outcomes?**
Presenter: **Francesca Bovis**, University of Genoa, Italy

Composite outcomes were proposed to study disability progression in MS, defined as the progression in at least one of a set of clinical variables as EDSS, T25FW, 9HPT and SDMT. Composite outcomes suffer from well-recognized limitations. Moreover, patients with different baseline demographic and clinical profiles can have a different propensity to respond to one specific domain. We combine the concept of defining responders to therapy according to their baseline profile with the concept of evaluating the treatment effect on multiple endpoints, with the idea that patients can have benefits from different outcomes. The treatment effect on 4 clinical endpoints (EDSS, T25FW, 9HPT, SDMT) was evaluated in an RCT assessing the Siponimod effect versus placebo. For each endpoint, a response score (RS) based on baseline characteristics was generated to characterize responders on the time to progression on that outcome, according to a previous method and using a training-validation procedure replicated on 500 bootstrap samples. Four different RS were obtained and validated, all showing a significant interaction with treatment, defining responders to each outcome. The scores were split into two groups according to a pre-specified cut-off. We showed that the treatment effect estimated in an RCT can be different on different outcomes and on different patients. This methodology allows personalizing the treatment effect according to the baseline patients' profile.

C0382: **Estimation of the marginal mean of recurrent events**
Presenter: **Giuliana Cortese**, University of Padua, Italy
Co-authors: Thomas Scheike

In survival analysis, currently, there is increasing interest in global summary measures over the follow-up period under study. When only a single event is of interest, examples of these measures are the mean lifetime and the residual mean lifetime. However, recurrent events such as cancer relapses or cardiovascular episodes can often be encountered in clinical and epidemiological studies on individuals who may potentially experience a terminal event such as death. With recurrent events data, a global summary measure of great interest is the marginal mean of the cumulative number of recurrent events experienced prior to the terminal event. Statistical efficiency of the IPCW nonparametric estimator for this mean is investigated and a novel efficient augmented estimator, based on dynamic predictions, is presented. In settings with different sources of heterogeneity, the proposed estimator is shown to improve efficiency greatly. In addition, regression models for the mean number of recurrent events can be employed to investigate the role of biomarkers on disease progression. A data example on chronic intestinal failures is discussed to highlight the practical use of these methods.

C0600: **Sample size and predictive performance of machine learning methods with survival data.**
Presenter: **Gabriele Infante**, University of Milan, Italy
Co-authors: Federico Ambrogi, Rosalba Miceli

Prediction models are increasingly developed and used in diagnostic and prognostic studies, where the use of Machine learning (ML) methods is becoming more and more popular over traditional regression techniques. For survival outcomes, the Cox proportional hazards model is largely used and it has been proven to achieve good prediction performances with few strong covariates. The possibility to improve the model performance by including non-linearities, covariate interactions and time-varying effects while controlling for overfitting must be carefully considered during the model building phase. On the other hand, ML techniques are able to learn complexities from data at the cost of hyper-parameter tuning and interpretability. One aspect of special interest is the sample size needed for developing a survival prediction model. While there is guidance when using traditional statistical models, the same does not apply when using ML techniques. A time-to-event simulation framework is developed to evaluate the performance of the Cox regression compared, among others, to tuned Random Survival Forest, Gradient Boosting and Neural Networks at varying sample sizes. We used simulations based on replications of subjects from publicly available databases, where event times were simulated according to a Cox model with non-linearities on continuous variables and time-varying effects. The SEER registry data were used for comparison with real-world data.

CO146 Room Aula Q IFCS SESSION: ASSESSMENT OF CLUSTER STABILITY AND PHYLOGENETIC INFERENCE Chair: Berthold Lausen

C0443: **Using aggregated cluster validity indexes to cluster football players performance data**
Presenter: **Christian Hennig**, University of Bologna, Italy
Co-authors: Serhat Akhanli

In cluster analysis applications, it is often difficult to decide between the many available clustering methods and to choose an appropriate number of clusters. We provide a case study to apply an approach based on several validation criteria that refer to different desirable characteristics of clustering, including stability. These characteristics are chosen based on the aim of clustering, and this allows the definition of a suitable validation index as a weighted average of calibrated individual indexes measuring the desirable features. We analyse football (soccer) player performance data with mixed type variables from the 2014-15 season of eight European major leagues. We cluster these data based on a tailor-made dissimilarity measure. We derive two different clusterings, namely a partition of the data set into major groups of essentially different players, and a second one that divides the data set into many small clusters, which can be used for finding players with a very similar profile to a given player. It is discussed what characteristics are desirable for these clusterings. Weighting the criteria for the second clustering is informed by a survey of football experts.

C0469: **Phylogeny and artificial neural networks**
Presenter: **Alina F Leuchtenberger**, Medical University of Vienna, Austria
Co-authors: Stephen M Crotty, Tamara Drucks, Heiko A Schmidt, Sebastian Burgstaller-Muehlbacher, Arndt von Haeseler

In recent years Artificial Neural Networks (ANNs) have become extremely popular. As powerful learning methods, they solve pattern recognition tasks and other challenges. We demonstrate how ANNs can be employed to solve phylogenetic problems. Long-branch attraction is a classical problem in phylogenetics. When long branches are placed adjacent to each other on a reconstructed tree, it is difficult to tell if this is artefactual (Felsenstein-type), or accurate (Farris-type). We developed F-zoneNN, an ANN which is able to infer with high accuracy if a multiple sequence alignment evolved under a Farris-type or Felsenstein-type tree. Despite its success, it is difficult to identify the features within the data that F-zoneNN leverages in making its determination. F-zoneNN is a composition of 9 linear and 9 non-linear functions including more than 1.2 million parameters, and so it is impossible to tell what drives the decisions of such an ANN. To get deeper insights into the decision-making process we endeavoured to simplify our trained network as much as possible, without sacrificing accuracy. This led to the development of an alternative mathematical representation of sequence alignments. Using this representation as input, we found that a linear function can infer the tree-type with high accuracy. This technique harbours the great potential for use in other phylogenetic applications of ANNs.

C0676: **Parametric bootstrap evaluation of unsupervised statistical learning and applications**
Presenter: **Berthold Lausen**, University of Essex, United Kingdom

Ultrametrics and additive tree metrics are mathematical models of phylogenetic inference or hierarchical clustering of distance data, which can be seen as unsupervised statistical learning problems. An additive measurement error model for distance data was used previously to develop a three objects variance estimator which provides a point estimate of the variance parameter without estimating the overall phylogenetic tree as an ultrametric or additive tree. Estimating the unknown location parameter, ultrametric or dendrogram, the three-objects variance estimator is used to compute parametric bootstrap estimates of the probability to observe the estimated clusters. The approach is applied in the context of user segmentation based on online behavioural data and compared to other recent suggestions for the evaluation of unsupervised statistical learning.

CO109 Room Aula E DYNAMIC MODELS FOR DISCRETE TIME SERIES AND LONGITUDINAL DATA Chair: Roberto Di Mari

C0377: **Extending the Poisson hidden Markov model to the multilevel framework with individual random effects**
Presenter: **Sebastian Mildiner Moraga**, Utrecht University, Netherlands
Co-authors: Emmeke Aarts

Hidden Markov models (HMMs) are probabilistic methods in which observations are seen as realizations of a latent Markov process with discrete states that switch over time. Moving beyond standard statistical tests, HMMs offer a statistical environment to optimally exploit the information present in multivariate time series, uncovering the latent dynamics that rule them. Although many applications of the HMM to model multivariate count data exist, so far, the support for multilevel data is restricted to non-parametric discrete random effects that apply to groups of individuals. We extend the Poisson HMM to the multilevel framework, accommodating variability between individuals with continuously distributed individual random effects, and we describe how to estimate individual and group-level parameters in a fully parametric Bayesian approach. The proposed model allows for probabilistic decoding of the sequence of hidden states based on individual-specific parameters and multivariate count time-series, and offers a framework to measure between-individual variability formally. Finally, we illustrate how to use our model to explore the latent dynamics governing complex count data with an empirical data set of multi-electrode electrophysiological measurements in macaque monkeys and a small Monte Carlo simulation.

C0453: **Evaluating complex agency effects on status transitions: Challenges within a Latent Markov Model paradigm**
Presenter: **Marco Doretti**, University of Perugia, Italy
Co-authors: Giorgio eduardo Montanari, Francesco Bartolucci, Maria Francesca Marino

Among their various purposes, Latent Markov models (LMMs) can be useful tools to cluster and/or rank a set of agencies operating on different users, for which some categorical variables measuring an unobserved trait of interest are collected over time. To this end, extensions of the basic LMM have been proposed in order to incorporate agency effects either as fixed or random effects. The focus is on a specific setting where: i) agency evaluation involves effects on transition probabilities only, and ii) these effects have a complex structure that cannot be captured by a single component, sometimes due to the data collection mechanism in use. A suitable example is represented by the assessment of the performance of nursing homes with regard to their ability to avoid residents' health status worsening. Building upon the existing literature employing LMMs in this framework, some issues related to the construction of a proper performance measure (and of a measure of its variability) are analyzed.

C0351: **Bias-corrected robust estimation of dynamic panel data models**
Presenter: **Pavel Cizek**, Tilburg University, Netherlands

Panel data are increasingly used due to their wider availability. The so-called fixed-effect panel models are difficult to estimate in the presence of outliers especially when the lagged values of the dependent variable are included and the number of time periods is small or moderate. Except for the median-ratio estimator, only locally robust methods based on the generalized method of moments (GMM) adjusted to have a bounded influence function were studied. To design robust regression estimators with positive breakdown points, we first generalize some existing robust regression estimators to preserve their robust properties when applied in dynamic models. However similarly to other linear-regression estimators, the proposed robust-regression estimators exhibit bias when applied to the first-differenced panel data. To address this, we next derive their asymptotic biases to facilitate the bias-correction procedures. We analyze the applicability and robustness of the bias correction procedures based on the derived asymptotic bias and on the computational methods such as jackknife, bootstrap, and indirect inference.

CO180 Room Aula F COMPUTATIONAL STATISTICS FROM THE LENS OF YOUNG RESEARCHERS II Chair: Riccardo Ceccato

C0225: **A unifying framework for rank and pseudo-rank based inference using nonparametric confidence distributions**
Presenter: **Jonas Beck**, Paris-Lodron-University of Salzburg, Salzburg, Austria, Austria
Co-authors: Arne Bathke

Nonparametric confidence distributions estimate statistical functionals by a distribution function on the parameter space, instead of the classical point or interval estimators. The concept bears analogy to the Bayesian posterior, but is nevertheless a completely frequentist concept. In order to ensure the desired statistical properties, we require that the cumulative distribution function on the parameter space is evaluated at the true parameter, uniformly distributed over the unit interval. The main focus lies on developing confidence distributions for the nonparametric relative effect and some natural extensions thereof. We develop asymptotic, range preserving and – especially important in the case of small sample sizes – approximate confidence distributions based on rank and pseudo-rank procedures. Due to the close relationship between point estimators, confidence intervals and p-values, these can all be approached in a unified manner within the framework of confidence distributions. The main goal is to make the powerful theory of confidence distributions available in a nonparametric context, that is, for situations where methods relying on parametric assumptions are not justifiable. Application of the proposed methods and interpretation of the results is demonstrated using real data sets, including ordinal and non-metric data.

C0343: **Permutation tests for C-sample problems: A multivariate scenario**
Presenter: **Elena Barzizza**, Universita degli studi di Padova, Italy
Co-authors: Rosa Arboretti, Nicolo Biasetton, Marta Disegna, Luca Pegoraro, Luigi Salmaso

The comparison between multivariate populations, which are characterized by a large number of outcome variables V, can be very challenging and particularly interesting in case we have $c > 2$ samples (where c is the number of samples considered). In this context, it may be worth considering the application of the Nonparametric Combination (NPC) methodology. The NPC methodology is well known to be flexible and quite powerful in situations characterised by multivariate data and numerous samples. For this reason, we introduce and compare a couple of NPC-based tests to deal with a specific real-world problem, in which we need to order $c > 2$ products according to $V > 1$ performance measures (i.e. c multivariate samples need to be compared). The application of these testing procedures allows us to demonstrate the flexibility of the Nonparametric Combination methodology and how it can be adopted to address a really common industrial problem.

C0426: **A novel active learning criterion for experiments with multiple responses**
Presenter: **Luca Pegoraro**, University of Padova, Italy
Co-authors: Rosa Arboretti, Elena Barzizza, Nicolo Biasetton, Riccardo Ceccato, Marta Disegna, Luigi Salmaso

Active Learning (AL) is a branch of Machine Learning (ML) in which the learner is in charge of the choice of data from which to learn. Its concept interweaves with the ones of adaptive sampling and sequential design from the experimental design literature, as the objective is to sample those data points that maximize information in terms of an acquisition criterion. We present a novel criterion that can drive adaptive batch sequential acquisition when the objective is to maximize the predictive accuracy of the models globally. The novel criterion is based on a ranking procedure that ranks candidate observations with respect to the uncertainty of predictions, a quantification of feature importance and a clustering approach for grouping candidates with respect to their proximity in the design space. The method can be applied when multiple responses are investigated in the same physical experiment and the data is noisy. We show the effectiveness of the proposed procedure through a simulation study and a case study application.

CC223 Room Aula G FORECASTING Chair: Aldo Goia

C0563: **Forecasting cardiorespiratory hospitalizations from air pollution levels through artificial neural networks**
Presenter: **Andrea Bucci**, universita degli studi g d annunzio di chieti pescara, Italy
Co-authors: Luigi Ippoliti, Pasquale Valentini

Air pollution is one of the most threatening risk factors for human health conditions. In fact, the epidemiological literature has widely proved that exposure to high levels of air pollution is associated with an increase in mortality and cardiorespiratory hospitalizations. Understanding how and if peaks in air pollution levels are capable of anticipating hospitalizations or death counts can be of great interest for policymakers to define public health strategies. In this context, the aim is to investigate the predictability of hospitalizations by cardiorespiratory diseases in Italian Provinces through the levels of ambient air pollution, such as nitrogen dioxide, sulfur dioxide and particulate matter. Since such a relationship is neither linear nor easy to be predicted, we propose to use neural networks to obtain the predictions of cardiorespiratory hospitalizations. In fact, neural networks can approximate any linear or nonlinear relationship and it has been already shown how they provide accurate time series predictions. Furthermore, recent extensions of traditional neural networks also allow accounting for spatial effects, which is a well-known characteristic of environmental phenomena. We compare their predictive accuracy with traditional neural networks and traditional spatio-temporal models.

C0243: **On predicting growth factor of daily new cases data of COVID-19 epidemic in Italy using ARIMA-ANN hybrid model**
Presenter: **Samir Safi**, United Arab Emirates University, United Arab Emirates

The Auto Regressive Integrated Moving Average, ARIMA model, cannot capture the nonlinear patterns exhibited by the 2019 coronavirus COVID-19 in terms of daily growth factor of daily new cases data in Spain. As a result, Artificial Neural Networks (ANNs) model is commonly used to resolve problems with nonlinear estimation. Different models that include ARIMA, ANNs, seasonal decomposition of time series, and a combination of these three models, hybrid model, were proposed to forecast the Growth Factor of COVID-19. The aim is to provide forecasting insights and criteria to use similar time series data to predict the growth factor of COVID-19 and to select the most suitable forecasting model for forecasting purposes. The best forecasting model selected was compared using the forecasting assessment criterion known as RMSE and MAE. The results add to the growing body of literature that seeks to accurately forecast the spread of COVID-19 by combining multiple models used by other researchers. The results are useful because it provides an accurate forecast for the growth factor of the COVID-19 epidemic. The importance of appropriate forecasts for policymakers to enhance better decision making is underscored.

C0575: **A joint use of monitoring and forecasting methods to detect change points in daily hospitalizations**
Presenter: **Rossella Miglio**, University of Bologna, Italy
Co-authors: Giulia Roli, Michele Scagliarini

Surveillance of hospitalization trends is a crucial task in health care and life sciences, especially during periods of a health emergency, such as those occurred in the recent pandemic. A good surveillance system supports decisions towards an optimal allocation of human, technical and economic resources and allows the development of better and innovative health policies. Under this framework, change points analysis represents a useful statistical tool to monitor the temporal trends in hospitalization, able to detect changes in the evolution of a phenomenon by estimating the corresponding time location. In particular, change point algorithms, such as control charts, can discern substantive causes of directional variation, from other causes of variation, i.e. random fluctuations around a baseline trend. The prediction of a trend is analogously a challenging objective, as it anticipates a change and, thus, further supports better decision-making processes towards early and more effective solutions. Statistical literature provides a large set of proper methods able to fulfil this aim. The monitoring methods and the forecasting techniques of time series are combined into a unique set of statistical tools proposed to detect change points in the trend of daily hospitalizations and applied to the forecast of COVID hospitalization in the Emilia Romagna region.

CC230 Room Aula B STATISTICAL MODELLING AND INFERENCE Chair: Francisco Louzada

C0690: **The Bayesian discrepancy measure: A new method for Bayesian inference**
Presenter: **Mara Manca**, University of Cagliari, Italy
Co-authors: Francesco Bertolino, Silvia Columbu, Monica Musio

The aim is to construct an index that, in the Bayesian context, allows to check the conformity of a given hypothesis with respect to the available information (prior distribution and data). The proposed evidence measure, called Bayesian Discrepancy Measure (BDM), has properties of consistency and invariance. After presenting the BDM and the related Bayesian Discrepancy Test (BDT), we show their conceptual and interpretative simplicity that allows us to easily deal with complex case studies that have not yet been addressed in the literature. Theoretical and computational developments of the BDM in more general contexts, such as model selection are also at an advanced stage.

C0691: **Bayesian modeling of time series of counts under censoring**
Presenter: **Isabel Pereira**, University of Aveiro, Portugal

Censored time series arise when explicit limits are placed on the observed data and occur in several fields including environmental monitoring, economics, medical and social sciences. The censoring may be due to measuring device limitations, such as detection limits in air pollution or mineral concentration in water. Censoring may also occur when constraints or regulations are imposed, such as in international trade studies where exports and imports are subject to trade barriers or hours worked, often treated as censored variables. The time series of counts under censoring are considered, focusing on the Poisson first-order integer-valued autoregressive(PoINAR) models. This class, while being simple and flexible, is useful for modelling positive-valued and integer-valued time series possessing an autoregressive structure with a non-negative serial correlation. Two natural approaches are investigated to analyze censored PoINAR(1) time series under the Bayesian framework: the Approximate Bayesian Computation (ABC) methodology and the Gibbs sampler with Data Augmentation (GDA) Approach. The parameter estimation performance of both approaches is made through a simulation study.

C0696: **Detecting breaks in certain random intensities through sequential testing on point processes**
Presenter: **Moinak Bhaduri**, Bentley University, United States

As we surface, probably momentarily, from the pandemic, other crises thwart normalcy: gross inequality, climate calamity, distressed refugees, and upped possibilities of a fresh Cold War. The enduring motif of our time is constant chaos. Frequently, that chaos results when one type of stationary system gives way to another. Change detection is mainly about estimating these points of deviation. Suppose a Poisson-type point process carries the system forward. In that case, we will offer a brand of detection algorithms, engineered through permutations of trend switched statistics and a judicious application of false discovery rate control. Certain members of this family that remain asymptotically consistent and close to the ground truth (evidenced through some Hausdorff-similarity) are isolated from pinpointing estimated change locations. Efficient forecasting proves to be a natural result. Change point-based clustering tools will also be examined. We will describe how such analyses offer concrete definitions to vague objects like Covid waves and measure their enormity.

CC229 Room Aula C MISSING DATA Chair: Victor Hugo Lachos Davila

C0557: **Missing-data analysis with power M-estimators**
Presenter: **Gabriel Frahm**, Helmut Schmidt University, Germany

The problem of estimating the location and scatter of incomplete multivariate data has been treated previously. An observed-data M-estimation procedure has been developed, in contrast to the common methods of missing-data analysis, and the procedure has been evaluated in simulation studies. Nonetheless, a comparison between well-known methods of imputation and the observed-data technique, which is used by the authors, is still missing. Our main objective is to present a comprehensive simulation study that solves the open question of whether it is better to rely on imputation or not. This question is highly relevant since observations are rarely complete in real-life applications.

C0646: **An R Package for bias reduction with LogF(1,1) penalty under the MAR mechanism**
Presenter: **Muna AL-Shaaibi**, Sultan Qaboos University, Oman
Co-authors: Ronald Wesonga

When data are missing at random (MAR), bias in estimated logit model parameters is inevitable. Although most studies proceed to apply and even publish such results, they are usually misleading and get worse with an increased proportion of missingness. We propose an R package for the bias reduction method, originating from our current study, which explores the penalization of the log-likelihood with the LogF(1,1) penalty. The four main functions with different purposes, including; filling the missing data, applying the Expectation-Maximization (EM) by the method of weights for the missing data, fitting the penalized binomial model under missingness, and summarizing the model output will be presented. The package has been validated using real-life COVID-19 data and some results are discussed.

C0654: **Improving the power of a test for detecting "missing not at random"**
Presenter: **Jack Noonan**, Cardiff University, United Kingdom
Co-authors: Robin Mitra, Stefanie Biedermann

Missing data is known to be an inherent and pervasive problem in the process of data collection. The effects are wide-ranging and the loss of data can lead to inefficiencies and introduce bias into analyses. The specific problem of data missing not at random (MNAR) is known to be one of the most complex and challenging problems to handle in this area and testing its prevalence is of great importance. The presence of MNAR missingness can only be tested using a follow-up sample of the missing observations and therefore recovering a proportion of missing values in an efficient way could be crucial in saving the experimenter's costs and time and may result in new treatments/technology reaching the public faster. We develop a strategy to allow researchers to be in a position to be well informed about whether MNAR is a credible issue. Within a multiple regression setting, we demonstrate a proof of concept example and provide recommendations for how the follow-up sample of missing observations should be designed.

CC217 Room Aula I MIXED MODELS AND APPLICATIONS Chair: Domingo Morales

C0223: **Nesting random effects factors in fixed effects factors**
Presenter: **Dario Ferreira**, University of Beira Interior, Portugal
Co-authors: Sandra Ferreira, Celia Nunes, Joao Mexia

Balanced fixed effects models are extended into mixed models, obtaining an enrichment of these models, without increasing the number of required observations. We start by obtaining a least square like estimator of the variance components, which through its determination coefficient gives a validation, or not, of the model. Next, we carry out inference on the estimable vectors. Then, we consider the special case in which the mean and the vector of residues are independent.

C0224: **A simulation study considering mixed linear models with cumulants generated by a Weibull distribution**
Presenter: **Sandra Ferreira**, University of Beira Interior, Covilha, Portugal
Co-authors: Dario Ferreira, Celia Nunes, Joao Mexia

Linear mixed models are increasingly being used to answer practical problems in several research areas. We consider the case in which the components of the random part of a linear mixed model follow a Weibull distribution. We obtain the moments and cumulants of this distribution and study how to derive their parameters. A simulation study is presented and discussed in some detail.

C0585: **ML pipeline for radiomics-based survival analysis on CT images of patients with hepatic CRC metastases**

Presenter: **Anna Theresa Stueber**, Ludwig-Maximilians-Universitaet (LMU) Muenchen, Germany

Co-authors: Stefan Coors, Katharina Jeblick, Andreas Mittermeier, Osman Oecal, Balthasar Schachtner, Philipp Wesp, Max Seidensticker, Michael Ingrisch

Using a statistically rigorous approach to compare the prognostic performance of different machine learning (ML) configurations and feature sources (clinical data (cd), liver and tumor radiomics from CT images) for survival analysis in patients with hepatic metastases. Prospectively CT images and cd from 431 patients with hepatic metastases of colorectal carcinoma were analyzed. Liver and tumor metastases were segmented automatically (nnU-net) and 1218 radiomics features (rad-liver, rad-tumor) were calculated. A large-scale ML benchmark-pipeline for survival/risk prediction consisting of preprocessing, feature selection, dimensionality reduction (PCA), hyperparameter tuning and training of different models - elastic-net regression, random survival forest (RSF) and gradient boosting - was developed and evaluated via 10-fold cross-validation (CV) using the metric integrated Brier-score (IBS). Addressing dependency structures in the setup, a mixed-model approach was used to compare algorithm and data configurations. 60 ML pipeline configurations were evaluated, showing RSF performs constantly equal or better than the other two algorithms with best/lowest IBS value when tuned, without PCA using cd+rad-tumor features with mean IBS 0.167 (95%-CI: 0.158; 0.176). We investigated our comprehensive benchmark pipeline via a mixed-model evaluation for the optimization of radiomics-based risk prediction. Optimal prognostic performance was achieved with a tuned RSF on rad-tumor with cd.

Wednesday 24.08.2022 09:00 - 10:30 Parallel Session F – COMPSTAT2022

CV191 Room Aula B SEMI- AND NONPARAMETRIC METHODS (VIRTUAL) Chair: Dennis Dobler

C0620: **Nonparametric distribution estimators of sample maximum in iid settings**
Presenter: **Moriyama Taku**, Tottori University, Japan
Extreme value theory has constructed asymptotic properties of the sample maximum under some parametric assumptions. The focus is on the probability distribution estimation of the sample maximum. The traditional approach is parametric fitting to the limiting distribution - the generalized extreme value distribution; however, the model in finite cases is misspecified to a certain extent. We propose nonparametric estimators that do not need model specification. Asymptotic properties of the distribution estimators are derived. The numerical performances of the parametric estimator and the nonparametric estimators are compared. A simulation experiment demonstrates the obtained asymptotic convergence rates and clarifies the influence of misspecification in the context of probability distribution estimation. It is assumed that the underlying distribution of the original sample belongs to one of the Hall class, the Weibull class and the bounded class, whose types of the limiting distributions are all different: the Frechet, Gumbel and Weibull. It is proven that the convergence rate of the parametric fitting estimator depends on both the tail index and the second-order parameter and gets slow as the tail index tends to zero. The simulation results are generally consistent with the obtained convergence rates. Finally, we report two real case studies: the Potomac River peak stream flow data and the Danish Fire Insurance data.

C0625: **Two-sample modified Anderson-Darling test and its properties**
Presenter: **Masato Kitani**, Tokyo University of Science, Japan
Co-authors: Yuyan Ma, Hidetoshi Murakami
Two-sample testing problem is widely used in many scientific fields. We are interested in testing whether two independent samples of distributions are equal. One of the powerful and useful conventional tests for such a situation is the two-sample Anderson-Darling test. Many researchers have discussed various improvements to the Anderson-Darling test statistic. We propose two modifications of the Anderson-Darling test statistic that emphasize the upper or lower tails of the distribution. It is necessary to derive the limiting distribution of the modified Anderson-Darling test statistic for practical analysis. Then, we prove that the limiting distribution of proposed test statistics is expressed as the weighted sum of chi-squared random variables. Additionally, we investigate the convergence to the limiting distribution and compare the powers of proposed tests with existing tests for various distributions via simulation studies. Finally, the application to real datasets demonstrates the usefulness of the proposed tests.

C0273: **Inference for dependent error functional data with application to event related potentials**
Presenter: **Kun Huang**, Tsinghua University, China
Co-authors: Sijie Zheng, Lijian Yang
Estimation and testing is studied for functional data with temporally dependent errors, an interesting example of which is the event-related potential (ERP). B-spline estimators are formulated for individual smooth trajectories and their population mean as well. The mean estimator is shown to be oracally efficient in the sense that it is as efficient as the infeasible mean estimator if all trajectories had been fully observed without contamination of errors. The oracle efficiency entails asymptotically correct simultaneous confidence band (SCB) for the mean function, which is useful for making inference on the global shape of the mean. Extensive simulation experiments with various time series errors and functional principal components confirm the theoretical conclusions. For a moderate sized ERP data set, multiple comparisons is done by constructing paired SCBs among 4 different stimuli, over 3 components N450, N1, N2 separately or simultaneously, leading to interesting findings.

C0269: **Hypotheses testing of functional principal components**
Presenter: **Zening Song**, Tsinghua University, China
Co-authors: Lijian Yang, Yuanyuan Zhang
A procedure is proposed to test the hypothesis that the standardized functional principle components (FPCs) of a functional data are equal to a given set of orthonormal basis (e.g., the Fourier basis). Based on B-spline estimators of individual trajectories, a chi-square type statistic is constructed and shown to be oracally efficient in the sense that its limiting distribution is the same as an infeasible statistic if all unobserved trajectories were known by "oracle". The limiting distribution is shown to be an infinite Gaussian quadratic form, and a finite sample estimator of its quantile is shown to be consistent. A test statistic is proposed based on the chi-square type statistic and approximate quantile of the Gaussian quadratic form, which is shown to be asymptotically correct. Simulation studies are conducted to illustrate the finite performance of the proposed testing procedure. For an EEG (ElectroEncephalogram) data, the proposed procedure has confirmed an interesting discovery that the centered EEG data is generated from a small set of standard Fourier basis.

CI009 Room Aula G NON-REGULAR STATISTICAL ANALYTICS FOR NON-NORMAL DATA Chair: Tsung-I Lin

C0374: **Some skew distributions useful in model-based clustering**
Presenter: **Geoffrey McLachlan**, University of Queensland, Australia
Co-authors: Sharon Lee
The literature on non-normal model-based clustering has continued to grow in recent years. The non-normal models often take the form of a mixture of component densities that offer a high degree of flexibility in distributional shapes. They handle skewness in different ways, most typically by introducing latent skewing variable(s), while some others consider marginal transformations of the original variable(s). We focus on various scale mixtures of fundamental skew-symmetric distributions and methods for their fitting via the EM algorithm.

C0265: **skewlmm: An R Package for fitting skewed and heavy-tailed linear mixed models**
Presenter: **Victor Hugo Lachos Davila**, University of Connecticut, United States
Co-authors: Larissa Avila Matos, Fernanda Schumacher
Longitudinal data are commonly analyzed using linear mixed models, which, for mathematical convenience, usually assume that both random effect and error follow normal distributions. However, these restrictive assumptions may result in a lack of robustness against departures from the normal distribution and invalid statistical inferences. The R package skewlmm provides user-friendly tools to fit linear mixed models by considering the scale mixture of the skew-normal class of distributions, and this robust model formulation accounts for a possible within-subject serial dependence by considering some useful dependence structures, such as autoregressive order p (ARp) and damped exponential correlation (DEC). A real example is used to illustrate the methodology and software.

C0207: **High-dimensional generalized linear model for longitudinal data**
Presenter: **Mohammad Arashi**, Ferdowsi University of Mashhad, Iran
Co-authors: Mozhgan Taavoni
Variable selection and estimation are considered for the additive generalized mixed model with high-dimensional longitudinal data. We use spline approximation for the nonparametric components and use double-penalization for estimation. Under some regularity conditions, the oracle properties of the resulting estimators are established. We assess the performance of the proposed estimation strategy with some numerical analyses.

CO063 Room Aula C COPULA MODELS AND APPLICATIONS **Chair: Elisa Perrone**

C0534: **Copula modelling with penalised complexity priors**
Presenter: **Clara Grazian**, University of Sydney, Australia
Co-authors: Cristiano Villa, Liseo Brunero, Diego Battagliese

The use of penalised complexity priors is explored for assessing the dependence structure in a multivariate distribution. We use the copula representation and derive penalised complexity priors for the parameter governing the copula. We show that any alpha-divergence between a multivariate distribution and its counterpart with independent components does not depend on the marginal distribution of the components. This implies that the penalised complexity prior to the parameters of the copula can be elicited independently of the specific form of the marginal distributions. This represents a useful simplification in the model-building step and may offer a new perspective in the field of objective Bayesian methodology. We also consider strategies for minimising the role of subjective inputs in the prior elicitation step. Finally, we explore the use of penalised complexity priors in Bayesian hypothesis testing. Our prior is compared with competing default priors both for estimation purposes and testing.

C0471: **Quantifying directed dependence via dimension reduction**
Presenter: **Sebastian Fuchs**, University of Salzburg, Austria

A bivariate copula is defined that captures the scale-invariant extent of dependence of a single random variable Y on a set of potential explanatory random variables $X_1,...,X_d$. The copula itself contains the information on whether Y is completely dependent on $X_1,...,X_d$, and whether Y and $X_1,...,X_d$ are independent. Evaluating this copula uniformly along the diagonal, i.e., calculating Spearman's footrule, leads to the so-called 'simple measure of conditional dependence'. On the other hand, evaluating this copula uniformly over the unit square, i.e., calculating Spearman's rho, leads to a distribution-free coefficient of determination. We demonstrate the broad applicability of the above methodology in the context of feature selection and variable selection.

C0572: **Dependence parameters of some perturbation-based copulas**
Presenter: **Susanne Saminger-Platz**, Johannes Kepler University Linz, Austria
Co-authors: Anna Kolesarova, Adam Seliga, Radko Mesiar, Erich Peter Klement

A prominent example of a perturbation of the bivariate independence copula is the parametric family of Eyraud-Farlie-Gumbel-Morgenstern copulas modeling (only) small dependencies. Allowing a larger parameter range and truncating with the lower and upper Frechet-Hoeffding copula bounds leads to a comprehensive extension of the family as has been shown by Huerlimann. We introduce and discuss some perturbation-based bivariate copulas by involving flipping and truncation and discuss several of their dependence parameters.

C0283: **The evolution of poverty in the EU-28: a further look based on multivariate tail dependence**
Presenter: **Cesar Garcia-Gomez**, Universidad de Valladolid, Spain
Co-authors: Ana Perez Espartero, Mercedes Prieto-Alaiz

A graphical tool is proposed which is based on the copula function, namely the multivariate tail concentration function, to represent the dependence structure on the tails of a multivariate joint distribution. We illustrate the use of this function to measure dependence between poverty dimensions. In particular, we analyse how multivariate tail dependence between the dimensions of the AROPE rate evolved in the EU-28 between 2008 and 2018. We find evidence of lower tail dependence in all EU-28 countries, although this dependence is time-varying over the period analysed and the effect of the Great Recession on this dependence is not homogeneous over all countries.

CO069 Room Aula D INFERENCE FOR FUNCTIONAL DATA **Chair: Dominik Liebl**

C0391: **Robust detection for change-points in functional time series based on spatial signs and bootstrap**
Presenter: **Lea Wegner**, Otto-von-Guericke University Magdeburg, Germany
Co-authors: Martin Wendler

One main strategy in changepoint detection for time series is to project the data on a finite-dimensional space with techniques such as functional principal components. In contrast, there are recent proposals to base the statistical tests on the full functional information, typically modeled as Hilbert-space-valued time series. Up to now, test statistics for changepoint detection in functional time series are based on sample means and outliers can influence the test result. Generalizing the Wilcoxon statistic, we have constructed a new functional version of a two-sample U-statistic with a bounded antisymmetric kernel. We will present limit theorems for U-statistics with values in Hilbert spaces and deduce the asymptotic distribution of our changepoint statistic. Because of the boundedness of the kernel, the statistic is indeed robust against outliers. Since this class of test statistics does not rely on dimension reduction, the limit distribution provides an infinite-dimensional covariance operator as a parameter, which is difficult to estimate. Because of this, we propose a new variant of the dependent wild bootstrap adapted to U-statistics in Hilbert spaces.

C0334: **Confidence regions for the location of peaks of a smooth random field**
Presenter: **Samuel Davenport**, University of California, San Diego, United Kingdom

Local maxima of random processes are useful for finding important regions and are routinely used, in areas such as neuroimaging, for summarising features of interest. We provide confidence regions for the location of local maxima of the mean and standardized effect size (i.e. Cohen's d) given multiple realisations of a random process. We prove central limit theorems for the location of the maximum of mean and t-statistic random fields and use these to provide asymptotic confidence regions for peak mean and Cohen's d. Under the assumption of stationarity, we develop Monte Carlo confidence regions for peaks of the mean that have better finite sample coverage than regions derived based on classical asymptotic normality. We illustrate our methods on 1D MEG data and 2D fMRI data from the UK Biobank.

C0473: **Fair causal inference with functional data**
Presenter: **Tim Mensinger**, University of Bonn, Germany

Inference for functional treatment effects in linear function-on-scalar regression models is examined. Inference is accomplished using simultaneous confidence bands based on adaptive critical value functions that allow inference under fairness constraints, making the results interpretable both locally and globally. To leverage existing methods in causal inference, we extend the potential outcome framework, directed acyclic graphs, and robust variance estimation to the case of functional data. Our developments are motivated by a case study in sports biomechanics, where we seek to estimate the effects of forefoot running on ankle strength from observational data.

C0597: **Simultaneous inference with CoPE sets**
Presenter: **Fabian Telschow**, Humboldt University zu Berlin, Germany
Co-authors: Armin Schwartzman

Recently there has been an increased interest in statistical analysis of C(S)-valued random variables where S is a compact metric space. Different inference methodologies like COnfidence Probability Excursion (COPE)sets, Simultaneous Confidence Bands and statistical tests like relevant difference and bio-equivalence tests have been proposed and successfully applied. We show that all these concepts can be unified within the framework of COPE sets. In particular, we show thatCoPE sets can be used to construct relevant tube tests and equivalence tests.

CO059 Room Aula Q ADVANCES IN LATENT VARIABLE MODELS I (VIRTUAL) **Chair: Paolo Giordani**

C0312: **Causal mediation analysis with latent mediators and survival outcome**
Presenter: **Xinyuan Song**, Chinese University of Hong Kong, Hong Kong

A joint modeling approach is developed that incorporates latent traits into causal mediation analysis with multiple mediators and a survival outcome. A linear structural equation model is used to characterize the latent mediators with several highly correlated observable surrogates and depicts the relationships among multiple parallel or causally ordered mediators and the exposure. A proportional hazards model is used to derive the path-specific causal effects on the scale of hazard ratio under the counterfactual framework with a set of sequential ignorability assumptions. A Bayesian approach with Markov chain Monte Carlo algorithm is developed to perform an efficient estimation of the causal effects. Posterior propriety theory is established for the proportional hazards model with latent variables. The empirical performance of the proposed method is verified through simulation studies. The proposed model is then applied to a study on the Alzheimer's Disease Neuroimaging Initiative dataset to investigate the causal effects of the APOE-epsilon4 allele on the disease progression, either directly or through potential mediators, such as hippocampus atrophy, ventricle expansion, and cognitive impairment.

C0336: **A fused lasso penalization for the nominal response model**
Presenter: **Michela Battauz**, University of Udine, Italy

The nominal response model, which can be used to analyze the categorical responses to a set of items, does not assume a predetermined order for the response categories. Due to this feature, the model is particularly suitable to group the response categories, which is pursued through a fused lasso penalization. Besides forcing the slope parameters towards a common value, the penalty adopted shrinks these coefficients towards zero. This is particularly interesting in the multidimensional nominal response model, since it is able to perform the selection of the latent variables related to each item. Hence, the proposal tends to reduce the large number of parameters of this model either by grouping the slope parameters of the response categories, or by setting all the slope parameters of one latent variable to zero. Simulation studies show that the resulting estimator not only presents a lower root mean square error, but also has a lower bias in small samples than the maximum likelihood estimator.

C0359: **Joint sparse principal component analysis**
Presenter: **Katrijn Van Deun**, Tilburg University, Netherlands

Comparing multivariate relations between different groups forms the core of many studies in the empirical sciences. Latent variable approaches (e.g., principal component and factor analysis) are most useful to explore such multivariate relations. The loadings are key to the interpretation of these latent variable models as they express the strength of association of the observed variables with the latent variables. Preferably variables load on one or a few components only and have zero loadings elsewhere as this eases interpretation. In addition, when comparing multiple groups, also a clear distinction between those variables that function in the same way over groups and those that do not is needed: Loadings should be exactly equal between those groups where the variables function in the same way and unequal elsewhere. We propose a multigroup latent variable model, joint sparse principal component analysis, that has these properties. Sparsity is imposed using cardinality constraints while equal loadings are obtained as the result of a fusion penalty. We efficiently solve the estimation problem by using an alternating optimization procedure that includes the alternating direction method of multipliers as one of the steps and tunes the cardinality and fusion penalty with a BIC-like statistic.

C0499: **A misspecification test for hidden Markov models based on finite mixture models**
Presenter: **Silvia Pandolfi**, University of Perugia, Italy
Co-authors: Francesco Bartolucci, Fulvia Pennoni

In the context of longitudinal data, we investigate the relation between hidden Markov (HM) models and finite mixture (FM) models, to provide a misspecification test for the class of former ones. We show that an HM model may be seen as a particular FM model. Based on this idea, we develop a new class of FM models, denoted FM2 models, which is based on an augmented set of components and suitable constraints on the conditional response probabilities, given these components. We also derive conditions under which the two model formulations become equivalent, based on suitable constraints on the parameters of the FM2 model. Based on these results, we develop a likelihood ratio misspecification test for the latent structure of an HM model and a multiple version of this test, based on the Bonferroni correction for multiple tests, which may be used in presence of many latent states or time occasions. The proposal is studied by a series of simulations, aimed at assessing the performance of the proposed tests under different circumstances, and by a real data application, which also shows that the testing procedure may be used as a criterion for selecting the number of latent states of an HM model.

CC222 Room Aula H BIOSTATISTICS AND APPLICATIONS **Chair: Malgorzata Bogdan**

C0396: **A computer-aided diagnosis system to detect Parkinson disease by using acoustic features**
Presenter: **Carlos Javier Perez Sanchez**, University of Extremadura, Spain
Co-authors: Javier Carron, Yolanda Campos-Roca, Mario Madruga Escalona

Parkinson's Disease (PD) is a chronic progressive neurodegenerative disorder that has an impact on the patients' voice, among other symptoms. Early detection is key to improving the patients' quality of life. A computer-aided diagnosis system to detect PD has been developed. This system collects voice recordings from an app, extracts multiple relevant acoustic features, and uses a two-stage variable selection and classification method that has been specifically designed for this task. An experiment has been conducted to analyze the system performance. A total of 30 people affected by PD were recruited among voluntary members of the Regional Association for Parkinson's Disease of Extremadura (Spain). Also, 30 healthy subjects were recruited to approximately match age and sex. After obtaining the voice recordings, a feature extraction process is performed to provide 33 acoustic features. The numerical features fed 5 variable selection and classification methods, and the selected features and performance metrics were compared. The best accuracy rate of 92% was obtained under a stratified cross-validation framework. The same methodology was applied to a sex and age-matched subsample from an existing voice recording database (https://parkinsonmpower.org), leading to a considerably lower best accuracy rate of 71%. The results clearly show the importance of the voice recordings and the potential of the proposed system.

C0539: **Multivariate regression model and permutation MANOVA: Case study on mental health effects of covid-19 lockdown**
Presenter: **Michela Borghesi**, Universita degli Studi di Ferrara, Italy
Co-authors: Stefano Bonnini

The impact of a pandemic on a global scale is not only medical strictly speaking, but also psychological, economic and social, as the pandemic of Covid-19, has generated fear, panic and psycho-emotional mass effects. The risk of developing anxiety and depression has grown exponentially, as physical distancing measures are themselves a key risk for the mental health of individuals, who have developed a condition of loneliness over the months. The purpose concerns the analysis of the effects of some factors related to the covid-19 pandemic on the mental health of individuals during the lockdown period. Given the multidimensional nature of mental health, a multivariate regression analysis was carried out on sample data concerning a survey done by the University of Milano Bicocca in Italy. The goodness of fit of the model was tested through a permutation MANOVA based on the method of combined permutation tests.

C0664: **Statistical modeling of Health space based on metabolic stress and oxidative stress scores**
Presenter: **Taesung Park**, Seoul National University, Korea, South
Co-authors: Oran Kwon, Chanhee Lee

Lifestyle-related chronic diseases are heterogeneous and multifactorial. Accurate estimation and visualization of the current health state are essential to optimize health and alleviate the increasing burden on lifestyle-related chronic diseases. We proposed a health space for visualizing individuals' health status in multi-dimensional space based on scores of metabolic stress and oxidative stress. We considered three statistical models to build the health space and found the best model using Korea National Health And Nutrition Examination Survey data. We developed a quantitative measure to evaluate and compare three models. Through simulation studies, we confirmed that the proportional odds model showed the highest power in discriminating the health status of individuals. Further validation studies were conducted using two independent cohorts in South Korea. In the validation studies, we successfully demonstrated the usefulness of the proposed health space model.

C0454: **Multiblock analysis of mixed data with optimal scaling: Application in epidemiology**
Presenter: **Martin Paries**, Oniris, France
Co-authors: Evelyne Vigneau, Stephanie Bougeard

A common problem in modern science, especially in biology, is the exploration of the relationships between blocks of variables measured on the same observations. Unsupervised component-based multiblock methods, such as Generalized Canonical Correlation Analysis or Multiblock Principal Component Analysis, are well referenced and allow for exploring these relationships. However, they are designed for numeric data only and real data have actually various formats (i.e nominal, ordinal, numerical = mixed data). Several component-based methods are proposed in the literature to deal with mixed variables, the well-known ones pertaining to the framework of Optimal Scaling. More precisely, Optimal Scaling is based on the two-step ALSOS algorithm where the Optimal Scaling step (i.e quantification of variables) alternates with the least square estimation step of the model parameters. Within this Optimal Scaling context, we propose an exploratory multiblock method called Multiblock Principal Component Analysis with Optimal Scaling (MBPCAOS). The proposed MBPCAOS can mainly be compared to other ones in the same framework, such as MFAmix or Overals. The MBPCAOS method is illustrated in a real case study pertaining to epidemiology.

CC152 **Room Aula I** **ROBUST METHODS I** **Chair: Peter Filzmoser**

C0348: **Computational discussions within an integer time series setup using a novel Poisson-Lindley model**
Presenter: **Ane van der Merwe**, University of Pretoria, South Africa
Co-authors: JT Ferreira

A generalization of the Lindley distribution is proposed by allowing for a measure of noncentrality via a mixture approach. Essential structural properties are investigated and derived in explicit and tractable forms, and the estimability of the model is illustrated via real data. Subsequently, this model is used as a candidate for the parameter of a Poisson model, which allows for departure from the usual equidispersion restriction that the Poisson offers when modelling count data. This more robust Poisson-noncentral Lindley is also systematically investigated and characteristics are derived. The computational impact and value of these continuous- and discrete models are illustrated in both simulation studies as well as real data fittings. The discrete model is further illustrated within an integer autoregressive environment as the error choice, and the effect of the systematically-induced noncentrality parameter is investigated. The extended binomial thinning operator, as a generalized case of usual binomial thinning, is also implemented within this time series context for this previously unconsidered discrete model. This paves the way for future flexible modeling, not only as a stand-alone contender in Lindley-type scenarios but also in discrete time series scenarios when the often-assumed equidispersed assumption is not adhered to in practical data environments.

C0398: **A bootstrap comparison of robust regression estimators**
Presenter: **Patrik Janacek**, The Czech Academy of Sciences, Institute of Information Theory and Automation, Czech Republic
Co-authors: Jan Kalina

The least squares estimator in linear regression is well known to be highly vulnerable to the presence of outliers in the data. Available robust statistical estimators are preferable as alternatives to the classical least squares. It has been repeatedly recommended to use the least squares together with a robust estimator, where the latter is understood as a diagnostic tool for the former. In other words, only if the robust estimator yields a very different result, the user should investigate the dataset closer and search for explanations. This requires a formal hypothesis test. A bootstrap test of equality of two linear regression estimators is developed. Its performance is presented on several real economic datasets contaminated by outliers. Although robust estimation (and particularly the least weighted squares estimator) is beneficial in all these datasets, robust estimates turn out not to be significantly different from non-robust ones.

C0401: **All-in-one Robust Estimator of the Gaussian Mean**
Presenter: **Arshak Minasyan**, ENSAE, France
Co-authors: Arnak Dalayan

The goal is to show that a single robust estimator of the mean of a multivariate Gaussian distribution can enjoy five desirable properties. First, it is computationally tractable in the sense that it can be computed in a time, which is at most polynomial in dimension, sample size, and the logarithm of the inverse of the contamination rate. Second, it is equivariant translations, uniform scaling, and orthogonal transformations. Third, it has a high breakdown point equal to 0.5, and a nearly minimax rate breakdownpoint approximately equal to 0.28. Fourth, it is minimax rate optimal, up to a logarithmic factor, when data consists of independent observations corrupted by adversarially chosen outliers. Fifth, it is asymptotically efficient when the rate of contamination tends to zero. The estimator is obtained by an iterative reweighting approach. Each sample point is assigned a weight that is iteratively updated by solving a convex optimization problem. We also establish dimension-free nonasymptotic risk bound for the expected error of the proposed estimator. It is the first result of this kind in the literature and involves only the effective rank of the covariance matrix. Finally, we show that the obtained results can be extended to sub-Gaussian distributions, as well as to the cases of the unknown rate of contamination or unknown covariance matrix.

C0494: **Efficient computation of robust multivariate maximum association**
Presenter: **Pia Pfeiffer**, TU Wien, Austria
Co-authors: Andreas Alfons, Peter Filzmoser

Methods to measure association between multivariate datasets become increasingly important as more multimodal data is acquired. Canonical Correlation Analysis (CCA) is widely applied for this task but is neither robust in the presence of atypical observations nor well-defined in the high-dimensional case when more variables than samples are collected. Let R denote a bivariate measure of association. A measure of maximum association between two multivariate variables X and Y is defined via maximization of R between linear combinations of sets of variables: $\rho = \max_{||\alpha||=1,||\beta||=1} R(\alpha^T X, \beta^T Y)$. Using the Pearson correlation for the association measure R results in the first canonical correlation coefficient, while a robust choice of R yields a more robust estimator. These estimators have desirable theoretical properties, but computation can be a limiting factor: Methods that require the computation of covariance matrices, or are based on pairwise comparison, or grid-search do not scale well to high-dimensional data. We present an algorithm based on adaptive gradient descent and M-association derived from a bivariate M-scatter matrix for the computation of robust multivariate maximum association. Simulations illustrate the robustness properties of our approach, as well as its suitability for high-dimensional data. The presented algorithm can also be applied to other robust methods in the context of high-dimensional data analysis.

CC213 **Room Aula E** **MODEL-BASED CLUSTERING** **Chair: Francesco Sanna Passino**

C0440: **Modelling three-way RNA sequencing data using matrix-variate Gaussian mixture models**
Presenter: **Theresa Scharl**, Boku Vienna, Austria

Co-authors: Bettina Gruen

RNA sequencing of time-course experiments leads to three-way count data where the dimensions are the genes, the time points and the biological units. Clustering of RNA-seq data allows the detection of groups of co-expressed genes over time. After standardisation, the normalised counts of individual genes across time points and biological units constitute compositional data. We propose the following procedure to suitably cluster the standardised three-way RNA-seq data: (1) Transform the data using the isometric log-ratio transform to map the composition in the D-part Aitchison-simplex to a D-1 dimensional Euclidean vector and (2) analyse the transformed RNA-seq data using Gaussian mixture models. To account for the three-way structure, we suggest using matrix-variate Gaussian mixture models to find groups of genes with similar expression patterns over time by simultaneously taking into account the different process conditions. This enables the specification of more parsimonious models assuming suitable time or biological effects within and across clusters. Such models also allow for an easier interpretation of the fitted model and the clusters obtained. The proposed three-way clustering approach will be applied to RNA-seq data from E. coli bioproduction processes and also compared to the two-way approach after flattening out the biological units.

C0542: **Clustering longitudinal ordinal data**
Presenter: **Francesco Amato**, University Lyon II, France
Co-authors: Julien Jacques

In social sciences or medicine, studies are often based on questionnaires asking participants to express ordered responses several times over a study period. A model to perform temporal clustering on such longitudinal ordinal data is presented, by assuming that the observed ordinal data are realizations of underlying matrix-normal distributions. Thus, the model relies on a mixture of matrix-variate normal distributions, accounting for the within and between time-dependence structures simultaneously. It allows for possible extension in a mixed data context, to deal jointly with continuous and ordinal data (and possibly more). An EM algorithm for the model estimation is developed. Applications on synthetic and on real data are presented.

C0545: **A stochastic block model for hypergraphs**
Presenter: **Luca Brusa**, Universita di Milano Bicocca, Italy
Co-authors: Catherine Matias

Over the past few decades a broad variety of models has been developed for graphs. However, modern applications in various fields highlighted the need to account for higher-order interactions, to include the information deriving from groups of three or more nodes. Simple examples include group interactions in social networks, scientific co-authorship, interactions between more than two species in ecological models or high-order correlations between neurons in brain networks. Hypergraphs provide the most general formalization of higher-order interactions: similarly to a graph, a hypergraph is defined as a set of nodes and a set of hyperedges, the latter specifying nodes taking part in each interaction. We propose a stochastic block model for hypergraphs to perform model-based clustering, capturing the information deriving from higher-order interactions. The formulation is sufficiently flexible to account for possible simplified latent structures. A variational expectation-maximization algorithm is developed to perform parameter estimation and model selection is explored using the ICL criterion. The model is applied to both simulated and real data, and the performance of the proposal is assessed in terms of parameter estimation and ability to recover the clusters. The estimation algorithm was implemented in C++ language and it was made available for the R software.

C0503: **A dynamic latent block model for co-clustering of zero-inflated count data streams**
Presenter: **Giulia Marchello**, Universite Cote dAzur, Inria, France
Co-authors: Marco Corneli, Charles Bouveyron

The simultaneous clustering of observations and features of data sets (known as co-clustering) has recently emerged as a central machine learning application to summarize massive data sets. However, most existing models focus on continuous data in stationary scenarios, where cluster assignments do not evolve over time. A novel latent block model is introduced for the dynamic co-clustering of count data streams with high sparsity. To properly model this type of data, we assume that the observations follow a time and block-dependent mixture of zero-inflated Poisson distributions, which combines two independent processes: a dynamic mixture of Poisson distributions and a time-dependent sparsity process. To model and detect abrupt changes in the dynamics of both clusters' memberships and data sparsity, the mixing and sparsity proportions are modeled through systems of ordinary differential equations. The model inference relies on an original variational procedure whose maximization step trains recurrent neural networks in order to solve the dynamical systems. Numerical experiments on simulated data sets demonstrate the effectiveness of the proposed methodology.

CC218 Room Aula F DESIGN OF EXPERIMENTS — Chair: Frederick Kin Hing Phoa

C0268: **Pair-switching rerandomization**
Presenter: **Ke Zhu**, Tsinghua University, China
Co-authors: Hanzhong Liu

Rerandomization discards assignments with covariates unbalanced in the treatment and control groups to improve the estimation and inference efficiency. However, the acceptance-rejection sampling method used by rerandomization is computationally inefficient. As a result, it is time-consuming for rerandomization to draw numerous independent assignments, which are necessary for performing Fisher randomization tests. To address this problem, a pair-switching rerandomization method is proposed to draw balanced assignments more efficiently. Under pair-switching rerandomization, the unbiasedness and variance reduction of the difference-in-means estimator are obtained, and valid Fisher randomization tests are developed. The proposed method is applicable in both non-sequentially and sequentially randomized experiments. Moreover, an exact approach is proposed to invert Fisher randomization tests to confidence intervals, which is faster than the existing methods and applicable to any experimental design. Comprehensive simulation studies are conducted to compare the finite-sample performances of the proposed method and classical rerandomization. Simulation results indicate that pair-switching rerandomization leads to comparable power of Fisher randomization tests and is 3-23 times faster than classical rerandomization. Finally, the pair-switching rerandomization method is applied to analyze two clinical trial data sets, both demonstrating the advantages of the proposed method.

C0274: **Constructing optimal order-of-addition designs using latin squares**
Presenter: **Shin-Fu Tsai**, National Taiwan University, Taiwan

Efficient order-of-addition designs can be very useful to study the impact of varying addition orders of several components in some industrial, chemical and pharmaceutical processes. We will introduce a systematic method to construct optimal order-of-addition designs. Based on the pairwise order model, a series of optimal and near-optimal designs can be easily generated by juxtaposing several isotopic Latin squares. First, an exchange algorithm will be introduced to search for efficient designs by performing column permutation on cyclic Latin squares. Next, when a small optimal design is generated by the proposed algorithm, a large optimal design can be obtained using a recursive method. By combining these approaches, many new optimal designs can be generated for real-world applications. The proposed designs are further compared with competing designs generated by conventional software. The results show that the proposed designs are often more efficient for estimating unknown parameters.

C0397: **Optimal design to test for heteroscedasticity in a regression model**
Presenter: **Samantha Leorato**, University of Milan, Italy
Co-authors: Chiara Tommasi, Alessandro Lanteri, Jesus Lopez-Fidalgo

The goal is to design an experiment to detect a specific kind of heteroscedasticity in a non-linear regression model, i.e. $y_i = \eta(x_i;\beta) + \varepsilon_i$, $\varepsilon_i \sim N(0;\sigma^2 h(x_i;\gamma))$, $i = 1,\dots,n$, where $\eta(x_i;\beta)$ is a possibly non-linear mean function, depending on a vector of regression coefficients $\beta \in \mathbb{R}^p$, and $\sigma^2 h(x_i;\gamma)$ is the error variance depending on an unknown constant σ^2 and on a continuous positive function h, completely known except for the parameter vector $\gamma \in \mathbb{R}^s$ and satisfies $h(\cdot;0) = 1$. We consider the testing problem $H_0 : \gamma = 0$ against a local alternative $H_1 : \gamma = \lambda/\sqrt{n}$, for some $\lambda \neq 0$ and n the sample size. The application of a likelihood-based test is a common approach to this problem, since its asymptotic distribution is known. The aim consists in designing an experiment with the goal of maximizing (in some sense) the asymptotic power of a likelihood-based test. Few papers in optimal design of experiments are related to hypothesis testing most of which concern designing to check an adequate fit to the true mean function. We justify the use of the D_s−criterion and the KL-optimality to design an experiment with the inferential goal of checking for heteroscedasticity.

C0479: **A comparative study of methods for constructing optimal designs**
Presenter: **Akram Mahmoudi**, Jonkoping International Business School, Sweden
Co-authors: Saumen Mandal

Owing to the widespread application of optimal designs, the motivation is to study some methods of optimum designs. Therefore, some methods are studied to construct approximate optimal designs. Some of these methods are gradient-based algorithms while some others are gradient-free algorithms. A comprehensive simulation study is done to study and compare the performance of the aforementioned methods. In particular, a rigorous study of a class of multiplicative algorithms is made. Then by using the properties of the directional derivatives of the criterion under study, attempts are done to improve convergence. Further, these methods are extended, and some strategies for constructing optimal designs are developed. Finally, a practical model in chemistry will be considered to illustrate the performance of the proposed methods.

CP001 Room Virtual Posters Room I POSTER SESSION I **Chair: Cristian Gatu**

C0439: **Forecasting implied volatility forecast by a parallel combination of CNN-bidirectional LSTMs**
Presenter: **DongWan Shin**, Ewha Womans University, Korea, South
Co-authors: Ji Eun Choi

A new forecast method is proposed based on artificial neural networks (ANNs), ensemble CNN-BiLSTM, which is an ensemble of three CNN-BiLSTMs constructed with the combination of Convolution Neural Network (CNN) and Bidirectional Long Short-Term Memory (BiLSTM). The new forecast method effectively handles the strong long memory serial dependence feature of the daily VXN by the ensemble CNN-BiLSTM together with proper normalization and batch size. The long memory features arising from time-dependent mean and variance is largely reduced by normalizing the data with local mean and local standard deviation (sd). The batch size is determined by the optimal block length of the moving block bootstrap which reflects the long memory. The ensemble CNN-BiLSTM concentrates on one-day, one-week and two-week features of the normalized VXN data. An out-of-sample forecast comparison reveals that (i) the proposed ensemble CNN-BiLSTM has better forecast performance than the autoregressive model, DNN, LSTM, BiLSTM, and individual CNN-BiLSTMs; (ii) the local mean-sd normalization has superior forecast performance to the standard global mean-sd normalization; (iii) the optimal block length improves the forecast performance over a batch size considered in the literature.

C0452: **Statistical network analysis for epilepsy MEG Data**
Presenter: **Jaehee Kim**, Duksung Womens University, Korea, South
Co-authors: Haeji Lee, Sunhan Shin

Network analysis is useful in understanding the structural and functional relationships between nodes by analyzing a mathematical model that represents a structure made up of a set of nodes and their connection forms. In neuroscience, network analysis using brain imaging data is actively being researched, but statistical methods for this are lacking. We intend to analyze statistical network analysis for epilepsy MEG data. The R program was used for the analysis, and the network was generated by using correlation coefficients of brain magnetic field signal data measured by 72 sensors. MEG data were collected from 44 Korean patients with epilepsy and 46 healthy controls. We compared the networks of healthy and patients including hub nodes, betweenness centrality, degree and network visualization. Time-varying networks were fitted using the TERGM(temporal exponential random graph model). The hub node in the patient group was a right posterior cingulate gyri node, while the hub node in the healthy group was a left anterior cingulate and paracingulate gyri node, both of which belong to the limbic system.

C0628: **Imputation for supervised learning problems in high dimension**
Presenter: **Hadrien Lorenzo**, INRIA, France
Co-authors: Jerome Saracco, Olivier Cloarec

The problem of missing data often occurs in data analysis. We consider missing values of the type MAR (Missing At Random). Then, the probability that a value is missing depends on one or multiple observed variables. Most modern algorithms focus on this type of missing values, and the most used implementations are certainly MICE, missForest, missMDA, or k-Nearest Neighbors imputations. To take into account sampling variability, it is better to propose M values for each missing value instead of a single one. This so-called "multiple imputation" procedure allows to provide proper imputation, in contrast to improper imputation. In practice, $M = 5$ is often sufficient. Most of the existing methods are not well suited to the high dimensional context, when the sample size n is much lower than the number of variables p, often symbolized as $n << p$. In supervised analysis, the variable y must be explained by the variable x. This implies that the part of x associated with y can be hard to find, when the classical imputation methodologies suffer. We present a new methodology, called Koh-Lanta, able to deal with missing values in a supervised context, using multiple imputation, and tackling the high dimensional issues. For the sake of simplicity, missing values are considered only in the x part.

C0254: **New developments on integral priors for Bayesian model selection**
Presenter: **Diego Salmeron Martinez**, Universidad de Murcia, Spain
Co-authors: Juan Antonio Cano Sanchez, Christian Robert

Integral priors were developed for Bayesian model selection and have been successfully applied in many situations. However, two aspects deserve special attention. First, the method is stated for the comparison of two models. Second, nonparametric density estimates of the integral priors have been typically needed to approximate the Bayes factors, which translates into more computing time. We generalize the definition for more than two models and propose new numerical procedures to approximate the Bayes factors. The method is illustrated with several examples, including location-scale models, Poisson versus the negative binomal family, hypothesis testing for the exponential distribution mean, and the problem of testing if the mean of the normal distribution with unknown variance is negative, zero, or positive. Finally, we illustrate the method for the variable selection problem.

C0418: **Covariance-based distributed fusion filter under random delays, packet dropout compensation and signal-noise correlation**
Presenter: **Raquel Caballero-Aguila**, Universidad de Jaen, Spain
Co-authors: Josefa Linares-Perez

The estimation problem in sensor networks has received considerable attention due to its multiple fields of application, which demand the development of new mathematical models and algorithms to accommodate the effect of unavoidable network-induced uncertainties. Especially significant are random transmission delays and packet dropouts, which, if not properly handled, may compromise the performance of estimators. The aim is to address the distributed fusion estimation problem of discrete-time stochastic signals from multisensor measurements subject to uncertainties modelled by random parameter matrices and additive noises, which are cross-correlated at the same time and correlated with the signal at the same and subsequent time steps. The signal evolution model is assumed to be unknown and only the mean and covariance functions of the processes involved in the measurement equations are available. Random one-step delays and packet dropouts occur during data transmission to the local processors and the prediction compensation strategy is used to attenuate the effect of packet dropouts. Using an innovation approach, a covariance-based least-squares recursive algorithm for the local filtering estimators is designed. Then, these local estimators are combined at a fusion centre to generate the proposed distributed filter as the least-squares matrix-weighted linear combination of the local ones.

C0436: **Using functional data analysis with electroencephalography data and introducing the `EEGLAB_FDA` MATLAB plugin**

Presenter: **Mohammad Fayaz**, Shahid Beheshti University of Medical Sciences, Iran

Electroencephalography (EEG) is widely used for medical research in neurosciences and medicine. In this regard, the EEGLAB is an open-source toolbox for the analysis of EEG dynamics trials. On the other hand, the functional data analysis (FDA) is a branch in statistics that is about the analysis of real-valued infinite dimension functions with different methods such as dimension reduction (ex, B-Spline, Functional Principal Component Analysis (FPCA)), functional canonical correlation analysis (FCCA), functional regression, and etc. The focus is on: firstly, providing a systematic review of the applications and statistical methods of the FDA with an EEG dataset that has 614 items in google scholar until the end of March 2022. The inclusion criteria are articles that were published as journal articles, English, and at least SCOPUS-indexed journals. Finally, 128 items (79 statistical methods and 49 non-statistical methods (with medical and clinical applications)) remain. A different aspect of the published articles such as journal names, types, subjects, publication years, FDA methodologies and etc. are provided. Second, we develop an open-source `EEGLAB_FDA` plug-in for EEGLAB in MATLAB that has some FDA graphical user interface (GUI) such as smoothing, FPCA, and FCCA. The first version is released and the next versions are coming.

C0470: **Study on nonparametric tests in linear model with autoregressive errors**

Presenter: **Martin Schindler**, Technical University of Liberec, Czech Republic

Co-authors: Olcay Arslan, Yesim Guney, Jana Jureckova, Jan Picek, Yetkin Tuac

In the linear regression model with possibly autoregressive errors, a family of nonparametric tests for regression under a nuisance autoregression is proposed. The tests are based on autoregression rank scores and avoid the estimation of nuisance parameters, in contrast to the tests proposed in the literature. The behavior of the proposed test is illustrated and its computation is described. In the simulation study, the power of the test is estimated under various settings of the parameters of the model.

Wednesday 24.08.2022 11:00 - 12:30 Parallel Session G – COMPSTAT2022

CV197 Room Aula Q STATISTICAL MODELLING AND INFERENCE (VIRTUAL) Chair: Fabrizio Durante

C0653: **Multiple non-crossing quantiles models for density forecasting**
Presenter: **Mauro Bernardi**, University of Padova, Italy
Co-authors: Francesco Lisi

Traditional quantile regression methods aim at modeling the conditional quantile of a response variable given a set of covariates and a given confidence level, thereby ignoring relevant information coming from adjacent quantiles. Information from multiple quantile estimates is usually combined "a posteriori" providing a complete picture of the conditional response. However, results of multiple comparisons of individual quantile estimates can be misleading whenever quantile curves cross each other. Recently, a few alternative approaches have been proposed to deal with the joint estimation of multiple quantile curves imposing non-crossing quantile conditions. We address this issue by taking advantage of a convenient Gaussian Markov random field (GMRF) representation of a penalty term acting on the multiple quantile loss function. In this way, we are able to straightforwardly incorporate dependence in the multiple quantile process. Theoretical properties of the process are investigated and an efficient algorithm for dealing with high-dimensional semi-parametric regression is provided. We show the effectiveness and the advantages of incorporating prior dependence into the multiple conditional quantile process by applying it to a data set of renewable energy productions with the goal of providing density forecasting measures of the production by means of semi-parametric additive models.

C0638: **Effect of censoring time on the statistical monitoring of lifetime data**
Presenter: **Chenglong Li**, Northwestern Polytechnical University, China

Life tests for highly-reliable products often take a long time even using accelerated life testing with censoring. When the production process is monitored by control charts with the lifetime as the key quality characteristic, the time spent on life testing could incur significant delays for the practitioners to make decisions after sampling. However, shortening the test duration, which results in excessive right-censored observations, inevitably degrades the test power for anomaly detection. Thus, close attention is paid to the determination of censoring time in life tests when monitoring lifetime data with likelihood-based control charts. To interpret the optimal censoring time, the performance metric, the out-of-control average time to signal (OC ATS), is deconstructed into two parts: the original OC ATS and the delay caused by life testing. Finite sample analytical and large sample asymptotic expressions of ATS are derived for type-I censored exponential lifetimes. Similar analytical expressions are also derived for the Weibull case. For general distributions, Monte-Carlo simulations are used for obtaining approximate results. Our numerical investigation uncovers the twofold influence of censoring time on the actual performance of control charts under various scenarios and provides useful references for practitioners to set more sensible censoring times in life testing.

C0647: **Bayesian semiparametric copula estimation and model selection: A comparison study**
Presenter: **Jichan Park**, Korea University, Seoul, Korea, Korea, South
Co-authors: Taeryon Choi

There have been developed a lot of different models using Copulas to model a multivariate probability distribution and develop an efficient inference. We propose an extended semiparametric bayesian bivariate copula model based on existing bayesian copula models. First of all, the dependence structure of bivariate random variables is modeled with various bivariate copulas, and mixtures of B-spline densities are used for marginal distributions. The proposed copulas proposed Student t copula, Clayton copula and Gumbel copula. They have the characteristic of maintaining the degree of dependence present in tails between random variables, so unlike Gaussian copula in the existed model, models with them can be appropriately used to explain data with different dependence structures. Also, we perform a bayesian model selection through a Reversible Jump Markov Chain Monte Carlo algorithm. In simulation studies of various settings, the performance of each model and the model selection were confirmed, and finally, a comparative analysis with real data sets was conducted using a stock price index and a crude oil data set.

C0661: **Constructing likelihood-ratio-based confidence intervals for multiple threshold parameters**
Presenter: **Luiggi Donayre**, University of Minnesota - Duluth, United States

A procedure to compute sequential (one-at-a-time) asymptotically-valid confidence intervals for n threshold parameters is proposed. Because the sequential procedure yields T-consistent estimates of the threshold parameters, the asymptotic distribution is identical to that for a model with multiple thresholds simultaneously estimated. Consequently, the limiting distribution of the likelihood ratio statistic, conditional on $(n-1)$ consistently estimated threshold parameters, is asymptotically free of nuisance parameters and critical values to construct confidence intervals for multiple threshold parameters can be derived. Using Monte Carlo simulations, conservative likelihood-ratio-based confidence intervals are shown to exhibit coverage rates at least as high as the nominal level for multiple threshold parameters, while only including relatively few observations of the threshold variable for each confidence interval.

CI011 Room Aula F MULTISTATE MODELS AND INTERMEDIATE EVENTS Chair: Martina Mittlboeck

C0203: **Analysis of survival data with multiple events: New contributions and practical recommendations**
Presenter: **Luis Machado**, University of Minho, Portugal

In many medical studies, patients may experience several events across a follow-up period. Analysis in such studies is often performed using multi-state models. These models can be successfully used for describing complex event history data, for example, describing stages in the disease progression of a patient. In these models one important goal is the modeling of transition rates but biomedical researchers are also interested in reporting interpretable results in a simple and summarized manner. These include estimates of predictive probabilities, such as the transition probabilities, occupation probabilities, cumulative incidence functions, prevalence and the sojourn time distributions. We aim to introduce feasible estimation methods for some of these quantities while providing some practical recommendations. The proposed methods are illustrated using real data. Software in the form of an R package developed by the authors will be introduced.

C0531: **Modelling the hazard of transition into the absorbing state in the illness-death survival model**
Presenter: **Laura Antolini**, Universita Milano Bicocca, Italy
Co-authors: Elena Tassistro, Davide Paolo Bernasconi, Paola Rebora, Maria Grazia Valsecchi

The illness death model is the simplest multi-state model where the transition from state 0 to the absorbing state 2 may involve an intermediate state 1. In this setting, the standard approach of analysis is the joint model of the transition hazards from 0 to 2 and from 1 to 2, including time to illness as a time-varying covariate and measuring time from origin even after the transition into state 1. The hazard from 1 to 2 can be also modelled i) separately on patients in state 1, measuring time from illness and including time to illness as a fixed covariate, or ii) by a joint model where time after the transition into state 1 is measured in both scales and time to illness is included as a time varying covariate. A further possibility is a joint model where time after the transition into state 1 is measured only from illness and time to illness is included as a fixed covariate. Through theoretical reasoning and simulation the use of these models will be discussed when aiming at:1) validating the properties of the illness-death process,2) estimating the impact of time to illness on the hazard from 1 to 2,3) quantifying the impact that the transition into 1 has on the hazard of the absorbing state.

C0554: **Evaluating longterm survival with generalised and weighted pseudovalues in paediatric stem cell transplantation studies**
Presenter: **Ulrike Poetschger**, Childrens Cancer Research Institute, Austria

Co-authors: Harald Heinzl, Martina Mittlboeck

Increasing the rate of long-term survivors is the common primary interest in paediatric oncology studies. In particular, the research question, of whether allogeneic stem-cell transplantation can improve long-term survival probabilities as compared to conventional chemotherapy, is statistically challenging. To avoid selection bias, the analysis has to be based on donor availability rather than actually performed stem-cell transplantation. Since donor search is ceased after a patient suffers an event, donor availability status is incompletely observed and a simple two-group comparison (genetic randomisation approach) is not possible anymore. Due to non-proportional hazards, the almost exclusively used Cox model with time-dependent covariates leads to ambiguous results and unreliable conclusions. Models based on generalised and weighted pseudo-values are novel statistical approaches to properly address this situation without relying on proportional hazards. The approaches allow the estimation and comparison of survival probabilities of patients with and without an available donor. Simulation studies demonstrate that the approaches are approximately unbiased, show satisfying confidence interval coverage, and are able to outperform other commonly used statistical techniques.

CO057 Room Aula B BAYESIAN TIME SERIES NOVELTY (VIRTUAL) **Chair: Michele Costola**

C0175: **Large vector autoregressions with stochastic volatility in mean**
Presenter: **Jamie Cross**, BI Norwegian Business School, Norway
Co-authors: Aubrey Poon, Gary Koop, Chenghan Hou

Recent research has shown that incorporating large datasets and volatility feedback effects into vector autoregression (VAR) models is useful for structural analysis. However, computational complexities have made it challenging to estimate VARs with both of these features. We propose an efficient Bayesian posterior simulator to estimate large VARs with stochastic volatility in mean (SVM) dynamics. The algorithm is more efficient (both computationally and statistically) compared to conventional particle-filter based algorithms. In two empirical applications on the US economy, the large VAR-SVM model provides (1) novel macroeconomic insights about multi-sectored spillovers of uncertainty and (2) competitive out-of-sample forecasts to conventional large VAR models.

C0444: **A general Bayesian approach to multiple-output quantile regression**
Presenter: **Annika Camehl**, Erasmus University Rotterdam, Netherlands
Co-authors: Dennis Fok, Kathrin Gruber

A general Bayesian approach is proposed to multiple-output quantile regression. We place a prior on the collection of regression functions and consider the joint distribution of the output-vector given the covariates to be a finite mixture of multivariate Gaussians. The resulting model provides an extremely flexible framework to approximate various symmetric and asymmetric distributions. We show how to construct conditional quantiles for the marginal effects within a unified model. It can handle correlation across disturbance terms, guarantees non-crossing quantiles and yields easily interpretable results.

C0468: **Model specification for Bayesian neural networks in macroeconomics**
Presenter: **Karin Klieber**, University of Salzburg, Austria
Co-authors: Florian Huber, Niko Hauzenberger, M. Marcellino

Relations in macroeconomic data are often nonlinear and subject to structural breaks. This is commonly captured through appropriate models. However, by choosing a specific model the researcher takes a strong stance on the nature and degree of nonlinearities. This gives rise to substantial model and specification uncertainty. We develop Bayesian neural networks that remain agnostic on the precise form of nonlinearities. Our model flexibly adjusts to the complexity of the dataset. This is achieved through Bayesian regularization techniques that adequately select the appropriate network structure without the necessity for using cross-validation. To investigate the degree and nature of nonlinearities in macroeconomic data, we train our neural network to four commonly used datasets in macroeconomics and finance. Our empirical results suggest that for cross-sectional data, a linear approximation works well in predictive terms whereas, for time series data, nonlinearities are important and especially so during turbulent times.

C0514: **Time-varying multilayer networks in a Bayesian spatial autoregressive model**
Presenter: **Michele Costola**, Ca' Foscari University of Venice, Italy
Co-authors: Matteo Iacopini, Casper Wichers

The impact of political relationships on financial stock markets is investigated. Political relationships among countries are inherently dynamic and characterized by multiple types (positive or negative) and varying intensity (strong or weak). To account for these data features, we propose a new spatial autoregressive model (SAR) by introducing time-varying multilayer networks. The model also incorporates country-specific network exposure and stochastic volatility to account for known stylized facts on financial data, together with a layer-specific parameter to capture the relative importance of each layer. We adopt a Bayesian approach to inference and design a new Markov Chain Monte Carlo algorithm based on Metropolis-Hastings and slice sampling to draw from the joint posterior distribution. The proposed method is used to investigate the impact of international political relationships among the top-GDP countries on their stock index returns. We find evidence of different network impacts across layers and countries.

CO097 Room Aula C SPORTS STATISTICS **Chair: Leonardo Egidi**

C0245: **G-RAPM: Revisiting players contributions in regularized adjusted plus-minus models for basketball analytics**
Presenter: **Luca Grassetti**, University of Udine, Italy

The estimation of the players' value based on on-field performances is a very relevant topic in the sports management framework. In fact, coaches and team managers can view ratings based on the players' evaluation manifold. Many approaches are adopted to determine the players' ratings. Among others, Regularized Adjusted Plus-Minus model can be considered one of the most interesting solutions. The model-based approach allows computing the efficiency of players by accounting for the opposing lineup effect and including some exogenous variables to mitigate the presence of confounding. The existing proposals suggest treating the players' effects as real-valued parameters defining contributions to the team's score. This solution is efficient and allows for direct comparison among players, but the interpretation of the players' performance measures is not always straightforward from the sports management's point of view. Following the well-known stochastic production frontier econometric approach, the proposed model specification considers players' contributions to the total lineup's score as positive constraint parameters. An additional unconstrained effect is introduced to adjust the total players' effects for the actual lineup contribution, which can identify both positive and negative synergies among individuals. The proposed model is estimated feasibly within the Bayesian framework, allowing straightforward further generalisations.

C0354: **Modeling and predicting the UEFA EURO 2020 with hybrid machine learning**
Presenter: **Andreas Groll**, Technical University Dortmund, Germany

Conventional approaches that analyze and predict the results of international matches in football are mostly based on the framework of Generalized Linear Models. The most frequently used type of regression model in the literature is the Poisson model. It has been shown that the predictive performance of such models can be improved by combining them with different regularization methods such as LASSO penalization. More recently, also methods from the machine learning field such as boosting and random forests turned out to be very powerful in the prediction of football match outcomes. We analyze both a hybrid random forest extension based on conditional inference trees and a hybrid boosting extension based on extreme gradient boosting for modeling football matches. The models are fitted to match data from previous UEFA European Championships (EUROs) and based on the corresponding estimates all match outcomes of the EURO 2020 are repeatedly simulated (100,000 times), resulting in winning probabilities for all participating national teams.

C0370: **Is the man-up situation really effective in women's waterpolo? A study on the 2020 European Championship**
Presenter: **Alessandro Lubisco**, University of Bologna, Italy

Women's team sports are becoming more and more interesting also from a technical point of view. In women's water polo, except for the slightly smaller playing area, the rules are the same as for men: seven players teams that compete for four quarters of 8 minutes of real play. After a major foul a player is sent out of play for 20 seconds. This is the so-called extra-man play action, thought to be very significant to the final result of a match. Teams with a higher percentage of goals in extra-man offence are more likely to win the game. For this reason, coaches dedicate a lot of time to training their team to attack and defend in this particular stage of the match. A study is performed into data from the 2020 European Women's Water Polo Championships, whose aim is to identify whether extra-man actions have any elements, like for example position where shots are executed or number and speed of passages, that lead to a goal or, at least, to a good shot, meaning a ball in the goal even if it is saved.

C0632: **Plus-minus a couple of millions: A machine learning model for transfer fee analysis**
Presenter: **Senthil Murugan Nagarajan**, University of Luxembourg, Luxembourg
Co-authors: Arne Maes, Dries Goossens, Lars Magnus Hvattum, Christophe Ley

One of the most popular sports is known to be Football, where team managers have major concerns for making important decisions about the player transfers, valuation-related issues, determination of market value, and transfer fees. Football clubs invest massive amounts of money in releasing or hiring players from clubs. However, it becomes a crucial task for club managers to estimate the value of a player in the transfer market. We propose various Machine Learning (ML) techniques, accompanied by feature selection methods, to build a model for predicting the players' transfer fees. More concretely, we used distinct regression techniques such as Lasso Regression, Ridge Regression, Random Forest Regression, Support Vector Regression, and Partial Least Square Regression, which we accompany with Interpretable ML techniques such as Variable Importance and Partial Dependence Plots in order to get insights into the most important predictors. The latter insight is particularly important for team managers.

CO170 Room Aula D VOLATILITY MODELS — Chair: Jean-Michel Zakoian

C0244: **Inference on multiplicative component GARCH without any small-order moment**
Presenter: **Baye Matar Kandji**, CREST/Institut Polytechnique de Paris, France
Co-authors: Christian Francq, Jean-Michel Zakoian

In multiplicative component GARCH models, the volatility is decomposed into the product of two factors which often receive interpretations in terms of "short run" (high frequency) and "long run" (low frequency) components. While two-component volatility models are widely used in applied works, some of their theoretical properties remain unexplored. We show that the strictly stationary solutions of such models do not admit any small-order finite moment, contrary to classical GARCH. It is shown that the strong consistency and the asymptotic normality of the Quasi-Maximum Likelihood estimator hold despite the absence of moments. Tests for the presence of long-run volatility relying on the asymptotic theory and a bootstrap procedure are proposed. Our results are illustrated via Monte Carlo experiments and real financial data.

C0278: **Linear regressions on time series**
Presenter: **Christian Francq**, CREST and University Lille III, France
Co-authors: Francisco Blasques, Sebastien Laurent

A score-driven autoregressive conditional beta (ACB) model is introduced that allows regressions with dynamic betas (or slope coefficients) and residuals with GARCH conditional volatility. The time-varying betas are allowed to depend on past shocks and exogenous variables. We establish the existence of a stationary solution for the ACB model, the invertibility of the score-driven filter for the time-varying betas, and the asymptotic properties of one-step and multi-step Gaussian QMLEs for the new ACB model. The finite sample properties of these estimators are studied by means of an extensive Monte Carlo study. Finally, we also propose a strategy to test for the constancy of the conditional betas. In a financial application, we find evidence for time-varying conditional betas and highlight the empirical relevance of the ACB model in a portfolio and risk management empirical exercise.

C0277: **A multivariate ARCH(∞) model with exogenous variables and dynamic conditional betas**
Presenter: **Julien Royer**, CREST, France
Co-authors: Christian Francq, Jean-Michel Zakoian

Factor models are highly common in the financial literature. Recent advances allow relaxing the constancy of slope coefficients (the so-called betas) by considering conditional regressions. The theory on the estimation of these dynamic conditional betas however usually relies on short memory volatility models, which can be restrictive in empirical applications. Moreover, exogenous variables have proven useful in recent studies on volatility modeling. We introduce a multivariate framework allowing for time-varying betas in which covolatilities can exhibit higher persistence than the standard exponential decay. Covariates are included in the dynamics of both conditional variances and betas. We establish stationarity conditions for the proposed model and prove the consistency and asymptotic normality of the QML estimator. Monte Carlo experiments are conducted to assess the performance of the estimation procedure in finite samples. Finally, we discuss the choice of potential relevant exogenous variables and illustrate the pertinence of the model on real data applications.

C0289: **An extended GARCH model with two volatility sequences**
Presenter: **Abdelhakim Aknouche**, Department of Mathematics, College of Science, Qassim University, Saudi Arabia
Co-authors: Christian Francq

A GARCH model with two volatility processes (2GARCH) is proposed. The first volatility, unobserved, satisfies an autoregressive conditional duration (ACD) equation based on past squared observations. The second observable (or predictable) volatility is exactly the same as that of a standard GARCH model and is none other than the conditional mean of the unobserved volatility. When the innovation of the ACD process degenerates at one, the two volatilities coincide and the 2GARCH model reduces to the standard GARCH model. Thus the ACD parameters are estimated in a standard way by using the Gaussian quasi-maximum likelihood estimate (QMLE), while the variance of the ACD innovation is consistently estimated by an appropriate weighted least squares method. The unobserved volatility sequence, however, can be estimated using signal extraction methods. Finally, the standard GARCH hypothesis is tested while testing the nullity of the variance of the ACD innovation. From the estimation of various return series (S&P 500, etc.), it turns out that the standard GARCH hypothesis cannot be reasonably accepted.

CO091 Room Aula I ADVANCES IN LATENT VARIABLE MODELS II (VIRTUAL) — Chair: Paolo Giordani

C0253: **Uncovering clusters and within-cluster variation in time series: Mixture multilevel vector-autoregressive modeling**
Presenter: **Anja Ernst**, University of Groningen, Netherlands
Co-authors: Marieke Timmerman, Feng Ji, Bertus Jeronimus, Casper Albers

In the social sciences, experience sampling methodology is increasingly used to analyze individuals' emotions, cognition and behaviors in everyday life. The objectives in the analysis of the resulting intensive longitudinal data are increasingly focused on inter-individual differences. To accommodate inter-individual differences to a great extent, a mixture of multilevel vector-autoregressive modeling is proposed. The model combines multilevel vector-autoregressive modeling with mixture modeling, to identify individuals with similar traits and dynamic processes.

This exploratory model identifies mixture components (or clusters) containing individuals with similar overall means, autoregressions, and cross-regressions. Within each component multilevel coefficients allow additionally for a within-component variation on these coefficients of interest. The model is illustrated on emotion data from the COGITO study. The COGITO data contains two samples of individuals from different age groups of over 100 individuals each. Participants' emotions were assessed daily for about 100 days. The advantage of exploratory identifying mixture components and accounting for within-component variation is illustrated in the COGITO data.

C0411: **lqmix: An R package to model longitudinal data via mixtures of linear quantile regressions**
Presenter: **Maria Francesca Marino**, University of Florence, Italy
Co-authors: Marco Alfo, Maria Giovanna Ranalli, Nicola Salvati

Quantile regression represents a well-established technique for the modeling of data when researchers are interested in the effect of predictors on the conditional quantiles of the response. When responses are repeatedly collected over time, dependence needs to be properly considered to avoid misleading inference. A standard way of proceeding is that of including unit-specific random coefficients in the model. The distribution of such parameters may be either specified parametrically or left unspecified. Following this latter approach, the lqmix R package for estimating the parameters of a linear quantile regression model for longitudinal data is introduced. Discrete, time-constant and/or time-varying, random coefficients are considered and estimated directly from the data, in a finite mixture perspective. Based on the nature of the random coefficients included in the model, a static, dynamic, or mixed-type mixture of linear quantile regression equations is obtained. An EM algorithm is used to derive estimates and a block-bootstrap procedure is employed for deriving model parameters' standard errors. Standard penalized likelihood criteria are used to identify the optimal number of mixture components. A potential missing at random mechanism in the responses is also taken into account.

C0429: **Approximate likelihood estimation of dynamic latent variable models for count data**
Presenter: **Silvia Cagnone**, University of Bologna, Italy
Co-authors: Silvia Bianconcini

When dynamic latent variable models are specified for discrete and or mixed observations, problems related to the integration of the likelihood function arise since analytical solutions do not exist. Our recently developed dimension-wise quadrature is applied to deal with these intractable likelihoods. A comparison is made with one of the most often used remedies discussed in the literature, which is the pairwise likelihood method. Both a real data application and a simulation study show the superior performance of the dimension-wise quadrature with respect to the pairwise likelihood in estimating the parameters of the latent autoregressive process.

C0509: **Fast and universal estimation of latent variable models using extended variational approximations**
Presenter: **Sara Taskinen**, University of Jyvaskyla, Finland
Co-authors: Pekka Korhonen, Francis Hui, Jenni Niku

Generalized linear latent variable models (GLLVMs) are a class of methods for analyzing multiresponse data. One of the main features of GLLVMs is their capacity to handle a variety of response types (e.g. counts, binomial and (semi-)continuous responses, and proportions data) as well as between response correlations. The inclusion of unobserved latent variables, however, poses a computational challenge, as the resulting marginal likelihood function involves an intractable integral. This has spurred research into approximation methods to overcome this integral, with a recent and particularly computationally scalable one being that of variational approximations (VA). Unfortunately, the closed-form variational lower bounds have only been obtained for certain combinations of response distributions and link functions. We thus propose an extended variational approximations (EVA) approach which widens the set of VA-applicable GLLVMs. In EVA we replace the complete-data likelihood function with its second order Taylor approximation about the mean of the variational distribution and obtain a closed-form approximation to the marginal likelihood for any response type and link function. We use simulation studies to demonstrate that EVA is competitive in terms of estimation and inferential performance relative to VA and a Laplace approximation approach, while being computationally more scalable.

CC212 Room Aula G DATA DEPTH **Chair: Stanislav Nagy**

C0381: **The influence function of scatter halfspace depth**
Presenter: **Germain Van Bever**, Universite de Namur, Belgium
Co-authors: Gaetan Louvet

Statistical depth provides robust nonparametric tools to analyze distributions. Depth functions indeed measure the adequacy of distributional parameters to underlying probability measures. In the location case, the celebrated (Tukey) halfspace depth has been widely studied and its robustness properties amply discussed. Recently, depth notions for scatter parameters have been defined and studied. The robustness properties of this latter depth function remain, however, largely unknown. We derive the influence function of scatter halfspace depth. Expressions are given in the known and unknown location case under mild distributional assumptions. In the latter case, the expression allows disentangling the unknown location effect from the scatter contamination. The corresponding asymptotic variance is also provided.

C0466: **Efficient computation of the angular halfspace depth**
Presenter: **Rainer Dyckerhoff**, University of Cologne, Germany
Co-authors: Stanislav Nagy, Petra Laketa

A great deal of research has recently focused on directional data, i.e., data on the unit sphere. The angular halfspace depth (also known as angular Tukey's depth) is a tool for non-parametric analysis of directional data. This depth was proposed already in 1987, but its widespread use in practice has been hampered by significant computational issues. An efficient algorithm is presented that is capable of exactly computing the angular halfspace depth in arbitrary dimension and that does not require the data to be in a general position. This algorithm is based on two projection schemes. In a first step, the data are repeatedly projected on a lower dimensional sphere. In a second step, the data are projected from the sphere to a linear space in which a variant of the usual halfspace depth is evaluated. Compared to the algorithm implemented in the R package 'depth', this new algorithm is considerably faster. A major advantage of the new algorithm is that the calculation of the depths of additional points with respect to the same dataset is extremely fast. In many cases, the calculation of 1000 depths requires less than ten times the time for calculating the depth of a single point.

C0530: **RKHS-based projection depths**
Presenter: **Arturo Castellanos**, Telecom Paris, France
Co-authors: Pavlo Mozharovskyi, Florence d Alche-Buc

Data depth is a statistical function that measures the centrality of an observation with respect to a distribution or a data set in a multivariate space. By exploiting the geometry of data, the depth function is fully non-parametric, robust, satisfies affine invariance, and is used in a variety of tasks as a generalisation of quantiles in higher dimensions. Despite its desirable statistical properties, data depth is often criticized - in particular among the machine learning community - for its inability to treat various types of data, high computational cost and difficulty to reflect multimodality of distribution. To improve on these aspects and unlock data depth computations for further types of data in a generic way, here, the data depth is defined in a reproducing kernel Hilbert space (RKHS) after asymmetrizing its univariate constituent. Further, due to the richness of the RKHS space, the search should be restricted to a properly chosen subspace. This approach allows us not only to better tackle data with multimodal or non-convex support, but as well to run an optimisation routine for depth computation. The appealing properties of this new class of depths are confirmed by a simulation study and a real-data benchmark.

C0577: **Approximate computation of projection depths**
Presenter: **Pavlo Mozharovskyi**, Telecom Paris, Institut Polytechnique de Paris, France
Co-authors: Rainer Dyckerhoff, Stanislav Nagy

Data depth is a concept in multivariate statistics that measures the centrality of a point in a given data cloud in a Euclidean space. If the depth of a point can be represented as the minimum of the depths with respect to all one-dimensional projections of the data, then the depth satisfies the so-called projection property. Such depths form an important class that includes many of the depths that have been proposed. For depths that satisfy the projection property, an approximate algorithm can easily be constructed since taking the minimum of the depths with respect to only a finite number of one-dimensional projections yields an upper bound for the depth with respect to the multivariate data. Such an algorithm is particularly useful if no exact algorithm exists or if the exact algorithm has a high computational complexity, as this is the case with the halfspace depth or the projection depth. To compute these depths in high dimensions, the use of an approximate algorithm with better complexity is surely preferable. Instead of focusing on a single method, we provide a comprehensive and fair comparison of several methods, both already described in the literature and original.

CC209 Room Aula H TEXT MINING **Chair: Peter Winker**

C0220: **Semiparametric latent topic modeling on consumer-generated corpora**
Presenter: **Dominic Dayta**, University of the Philippines, Philippines
Co-authors: Erniel Barrios

Legacy procedures for topic modelling have generally suffered overfitting problems and weakness in reconstructing sparse topic structures. SemiparTM, a two-step approach utilizing nonnegative matrix factorization and semiparametric regression in topic modeling, is proposed. SemiparTM enables the reconstruction of sparse topic structures in the corpus and provides a generative model for predicting topics in new documents entering the corpus. Assuming the presence of auxiliary information related to the topics, this approach performs better in discovering underlying topic structures in cases of corpora that are small and limited in vocabulary. In an actual consumer feedback corpus, SemiparTM also demonstrably provides interpretable and useful topic definitions comparable with those produced by the legacy methods.

C0316: **Gender differences in personality perceptions in the labor force: Use of new data sources**
Presenter: **Dania Eugenidis**, Justus-Liebig-University Giessen, Germany

Gender stereotypes still play a major role in the perception and representation of people in the workplace. Traditional measures, such as questionnaires, often lack objectivity and thus struggle to provide the full picture. However, evidence-based policymaking requires accurate indicators of gender inequalities to promote equality. This framework depicts the first ever study examining the external portrayal of gender stereotypes on a company level using publicly available big data. Specifically, over 1 million company websites are analysed using natural language processing. Shortcomings of traditional quantitative indicators are to be overcome regarding timeliness, granularity and cost efficiency. That way, it is possible for the first time to conduct fully automated, objective and almost comprehensive analysis of the linguistic portrayal of gender in a corporate context. Subsequent comparison to the literature takes place by contextualizing the gender stereotype measures following the personality traits of the Big Five-factor model and their sublevels. The results of the statistical analysis indicate significant stereotypes within personality traits for large portions of the sample. These differences in gender presentation are mostly consistent with those found in the literature, which serve as a validation for the presented framework.

C0618: **Testing the equality of topic distribution between documents of a corpus**
Presenter: **Louisa Kontoghiorghes**, Kings College London, United Kingdom
Co-authors: Ana Colubi

Topic modeling is a well-known text mining technique to identify the themes covered in a set of documents. We introduce two methodologies to test whether two documents of a given corpus are homogeneous with respect to the topics they cover. The suggested approach uses Latent Dirichlet Allocation (LDA) to estimate the topic distributions. Furthermore, Kullback-Leibler divergence and the chi-square are used separately to measure the distance between the distributions, and their results are compared. Since the sampling distribution of the proposed statistics is unknown, a (frequentist) bootstrap test is suggested. The methodology is illustrated using scientific abstracts from the CMStatistics conference.

C0388: **Difference in SDG reportings of research articles using zero-shot text classification**
Presenter: **Christoph Funk**, Justus-Liebig-University Giessen, Germany
Co-authors: Elena Toenjes, Lutz Breuer, Ramona Teuber

In September 2015, the United Nations (UN) set an agenda to transform our world by 2030 with the adoption of 17 Sustainable Development Goals (SDGs) and 169 targets and 231 indicators for monitoring. Since then, the academic literature on the SDGs has grown steadily. So far, it is not clear in which countries SDGs are predominantly addressed. We apply zero-shot classification as a text mining tool on SDG-related scientific articles to analyze the scientific discourse on the 17 SDGs. First, we review the scientific literature on the SDGs, which allows us to draw conclusions about the focal points of scientific discourse worldwide. Second, we show that abstracts contain the most relevant information from scientific articles related to the discussed SDGs. Third, we demonstrate that zero-shot text classification can be a useful tool to label extensive textual information, thus providing an efficient tool for policymakers to screen the scientific literature, but also to provide information beyond the typical UN indicators. Our results suggest that SDGs 1, 2, 4 and 5 are less likely to be discussed than the remaining 13 SDGs. We find considerable variations in the scientific discourse across countries worldwide. SDGs 1 and 3 show the most negative correlation between the likelihood of discussion and their indicators. In addition, SDGs 7, 9, 15 and 16 show a positive relationship and have a higher probability of being discussed, even if their indicators perform well.

CC208 Room Aula E CHANGE-POINT DETECTION **Chair: Berthold Lausen**

C0194: **Changepoint in dependent and non-stationary panels**
Presenter: **Michal Pesta**, Charles University, Czech Republic
Co-authors: Matus Maciak, Barbora Pestova

Detection procedures for a change in means of panel data are proposed. Unlike classical inference tools used for the changepoint analysis in the panel data framework, we allow for mutually dependent and generally non-stationary panels with an extremely short follow-up period. Two competitive self-normalized test statistics are employed, and their asymptotic properties are derived for a large number of available panels. The bootstrap extensions are introduced in order to handle such a universal setup. The novel change-point methods are able to detect a common break point even when the change occurs immediately after the first time point or just before the last observation period. The developed tests are proved to be consistent. Their empirical properties are investigated through a simulation study. The invented techniques are applied to option pricing.

C0492: **Change detection in dynamic networks using flexible multivariate control charts**
Presenter: **Jonathan Flossdorf**, TU Dortmund University, Germany
Co-authors: Carsten Jentsch, Roland Fried

The focus is on the identification of differences in dynamic networks, i.e. in a sequence of networks between various time points. This task is important for statistical procedures like two-sample tests or change-point detection. Due to the rather complex nature of dynamic network data, the complexity is typically reduced to a metric or some sort of a model based on these metrics. However, the reduction in network metrics can result in a heavy information loss. Hence, understanding their behaviour in various change scenarios is crucial. We present a categorization of different types of changes that can occur in dynamic network data. We analyze the suitability and limitations of common network metrics in such situations with respect to their mathematical properties and give comprehensive explanations of their behaviour. This leads to well-founded advice on which metrics to use in various application scenarios. Based on this foundation, we develop an online monitoring approach usable for flexible network structures and types of changes. It uses a sound choice of a set of the analyzed network metrics that are jointly monitored in a suitable multivariate control chart scheme, which performs superior to univariate analysis and enables both parametric and non-parametric usage. All our findings are supported by extensive simulation studies and real-world examples.

C0561: **Modelling and detecting changes in spatial time series**

Presenter: **Gaurav Agarwal**, Lancaster University, United Kingdom

Co-authors: Idris Eckley, Paul Fearnhead

Changepoints have been extensively studied for time series data, but there is limited literature on detecting changes in stochastic processes over time. A likelihood-based methodology is developed for the simultaneous estimation of both changepoints and model parameters of spatio-temporal processes. Contrasting to existing spatial changepoint methods, which fit a piecewise stationary model assuming independence across segments, we fit a nonstationary model without any independence assumption. To deal with the complexity of the full likelihood model, we propose a computationally efficient Markov approximation. We study the effect of such an approximation through a comprehensive set of simulations. Furthermore, we present a comparison with existing methodologies, both in the case of dependence and independence across segments. The method is employed for changepoint detection and missing data prediction in daily soil moisture concentrations across different sites in the United Kingdom over a period of two years.

C0496: **Changepoint detection in periodic behaviour**

Presenter: **Owen Li**, Lancaster University, United Kingdom

Co-authors: Rebecca Killick

Traditional changepoint approaches consider changepoints to occur linearly in time; one changepoint happens after another, and they are not linked. However, data processes may exhibit periodic behaviour and so changepoints will occur regularly, e.g. sleeping patterns and daily routine behaviour. Using linear changepoint approaches in these settings will miss global changepoint features which affect changepoints on the more local (periodic) level, for example, the introduction of local lock-downs affecting sleeping patterns. Being able to tease these global changepoint features from the more local (periodic) ones is beneficial for inference. We propose a deterministic periodic changepoint method using a periodic (circular) time perspective. This is done by adapting the Segment Neighbourhood changepoint method to the periodic time perspective. We then integrate this local changepoint model into the pruned exact linear time (PELT) search algorithm to identify the optimal global changepoint positions. We demonstrate that the method detects both local and global changepoints with high accuracy on simulations and motivating digital health applications.

Wednesday 24.08.2022 14:15 - 16:15 Parallel Session H – COMPSTAT2022

CV195 Room Aula B ALGORITHMS AND COMPUTATIONAL METHODS (VIRTUAL) Chair: Paolo Giordani

C0487: **Enumeration of substructures in convolution kernels for structured data: the case of the subtree kernel**
Presenter: **Romain Azais**, Inria, France
Co-authors: Florian Ingels

Kernel methods are particularly adapted to the analysis of complex combinatorial data (such as sequences, trees or graphs) which do not admit in general a Euclidean structure. Convolution kernels consist in defining the similarity between 2 data via the number of occurrences of certain substructures they share. In order to accurately compare data, one may want to choose a rich family of admissible substructures, but the difficulty then appears in the computation of the kernel: the richer the substructures involved, the more complex the computation of the kernel. In some cases, the kernel can be evaluated efficiently without having to enumerate the substructures in common. However, this can strongly constrain the parametrization of the weight function. Through the example of the subtree kernel, we state, both theoretically and numerically, the importance of the design of the weight function in supervised classification problems. We develop a new algorithm for computing the subtree kernel, based on the minimal enumeration of substructures by exact compression techniques of trees. This method allows to extract the important features and learn the weight function on the data. We establish the interest in this algorithm both in terms of complexity, prediction capability and data visualization, on 8 real databases. Finally, we show how more sophisticated enumeration techniques can be used to extend the kernel to higher orders and solve frequent pattern mining problems.

C0583: **Evaluating the effect of planned missing designs in the structural equation models fit measures**
Presenter: **Paula C R Vicente**, Lusofona University, Portugal

Missing or incomplete data represent a persistent problem in several studies. In a planned missing design, the non-responses occur according to the researcher's will, and the purpose of using such a design is to increase the quality of the data, avoiding the effort of inquiry. On the other hand, the estimation of a structural equation model consists in finding estimates for the model parameters that result in a variance-covariance matrix with the best fit to the theoretical model considered. There are different criteria to evaluate how well the theoretical model fits the observed data. The fit indices usually used in this type of modeling are the root mean square error of approximation (RMSEA), root mean square residual (SRMR), comparative fit index (CFI) and Tucker-Lewis Index (TLI). The aim is to explore the effect of the non-responses due to a planned missing design on the mentioned fit indices. A simulation study was conducted, considering different models, sample sizes, number of indicators, factor loadings and correlation between factors. For each simulated condition, 1000 replications were generated using the simsem package in R. Recommendations for best practices are discussed.

C0240: **Mean of exponential distributions: Estimation from sums of unequal size samples**
Presenter: **Miguel Casquilho**, FCIENCIAS-ID, Portugal
Co-authors: Jorge Buescu

The sums of unequal size samples are frequently measured and recorded in numerous industrial, manufacturing, and related activities, for Quality control reasons or supplier-customer security. While the also common, particular case of equal size samples is routine, the treatment of sums of unequal size samples seems to be absent in the literature. We address the estimation of the parameter of an adopted Exponential distribution of the individual items that underlie the samples, for the said circumstances where the sample sums are measured, thus, adding no further effort or cost to data collection. After having formerly solved the analogous problem for Gaussian items, we present the estimation of the single parameter of the Exponential, its mean, with point estimation and confidence intervals. The point estimation was derived by means of the Maximum Likelihood method. As in the Gaussian case, it gives a weighted average of the measured sample averages; and the confidence intervals are obtained computationally by Monte Carlo simulation. All these computations are freely available for direct use on our web pages, for proposed data or other supplied by the user.

C0311: **smoothEM: A new approach for the simultaneous assessment of smooth curves and spikes**
Presenter: **Marzia Cremona**, Universita Laval, Canada
Co-authors: Huy Dang, Francesca Chiaromonte

Many longitudinal data comprise both smooth and irregular elements. We consider scenarios in which an underlying smooth curve is composed not just of Gaussian errors, but also of irregular spikes that (a) are themselves of interest, and (b) can negatively affect our ability to characterize the underlying curve. We propose an approach that, combining regularized spline smoothing and an EM algorithm, allows us to both identify spikes and estimate the smooth component. We prove the convergence of EM estimates to the true population parameters under some assumptions. Next, we demonstrate the performance of our method on finite samples and its robustness to assumptions' violations through simulations. Finally, we apply it to the analysis of two time series on the annual heatwaves index in the US and on the weekly electricity consumption in Ireland. In both datasets, we are able to characterize underlying smooth trends and pinpoint irregular/extreme behaviors.

C0294: **On variance estimation in online problems**
Presenter: **Kin Wai Chan**, The Chinese University of Hong Kong, Hong Kong
Co-authors: Man Fung Leung

Online problems arise naturally in many fields of statistics. On top of them, modern computing allows intractable offline problems to be approached with online techniques. Nevertheless, variance estimation in online problems remains largely offline, which limits the practical value of inference-based techniques. We propose a general framework to construct efficient long-run variance estimators in online problems. The contributions lie in three aspects. Statistically, we derive the first set of sufficient conditions for $O(1)$-time or $O(1)$-space update, which allows our framework to generate online estimators that uniformly dominate existing alternatives. Computationally, we introduce mini-batch estimation to accelerate online estimators in practice. Implementation issues such as automatic optimal parameters selection are discussed. Practically, we demonstrate the possibility to use recursive (online and mini-batch) estimators in convergence diagnostics and learning rate tuning. We also illustrate the strength of our estimators in some standard online problems such as change-point detection and confidence interval construction.

CO029 Room Aula G DEPENDENCE MODELS Chair: Fabrizio Durante

C0177: **General comparison results for factor models**
Presenter: **Jonathan Ansari**, University of Freiburg, Germany
Co-authors: Ludger Ruschendorf

In the setting of a factor model, the components of a random vector $X = (X_1, ..., X_d)$ depend on a common factor variable Z and some individual influences. Typically, the marginal distributions, as well as the copula of X_i and Z, can be (partially) estimated from the data. However, the conditional distribution of X given Z is often not specified. Extending recently investigated ordering results for a multivariate version of *-products of bivariate copulas to the supermodular ordering, we provide some general comparison results for factor models in dependence on the specifications considering specifically the strong notion of the supermodular and directionally convex order. The proofs are based on standard classical ordering theory and rearrangement results, as well as on mass transfer theory. As a consequence of the ordering results, we derive best and worst case scenarios in relevant classes of factor models allowing, in particular, interesting applications to deriving sharp bounds in financial and insurance risk models. Further, we provide a new construction method for comprehensive, multi-parameter families of positively dependent, multivariate distributions that are increasing in the parameters w.r.t. the supermodular or directionally convex order.

C0296: **Non-central squared copulas: Properties and applications**
Presenter: **Bouchra Nasri**, University of Montreal, Canada

The goal is to introduce new families of multivariate copulas, extending the chi-square copulas, the Fisher copula, and squared copulas. The new families are constructed from existing copulas by first transforming their margins to standard Gaussian distributions, then transforming these variables into non-central chi square variables with one degree of freedom, and finally by considering the copula associated with these new variables. It is shown that by varying the non-centrality parameters, one can model non-monotonic dependence, and when one or many non-centrality parameters are outside a given hyperrectangle, then the copula is almost the same as the one when these parameters are infinite. For these new families, the tail behavior, and the monotonicity of dependence measures such as Kendall's tau and Spearman's rho are investigated. The estimation is discussed. Some examples will illustrate the usefulness of these new copula families.

C0639: **Extreme value copulas based on Freund's multivariate lifetime model**
Presenter: **Sandor Guzmics**, University of Vienna, Austria

The exponential distribution and its multivariate generalizations are widely used in lifetime modeling. There are numerous models that explicitly incorporate a dependence structure among the components, Freund's bivariate distribution is such one. Its copula has been presented previously. We also provided previously the corresponding bivariate extreme value copula and discussed a natural multivariate generalization of the model. The basic idea is that the remaining lifetime of any entity in a multivariate system is shortened when one of the other entities defaults. We investigate the dependence structure in Freund's multivariate lifetime model, assuming a symmetric parameter setting, i.e., when the initial lifetime intensities, as well as the shock parameters, are all the same. We present some remarkable properties of this multivariate, parametric copula family, and examine the corresponding extreme value copulas.

C0617: **Living on the edge: A unified approach to antithetic sampling**
Presenter: **Lorenzo Frattarolo**, European Commission Joint Research Centre (JRC), Italy
Co-authors: Roberto Casarin, Radu Craiu, Christian Robert

Recurrent ingredients in the antithetic sampling literature are identified which lead to a unified sampling framework. We introduce a new class of antithetic schemes that include the most used antithetic proposals. This perspective enables the derivation of new properties of the sampling schemes: i) optimality in the Kullback-Leibler sense; ii) closed-form multivariate Kendall's τ and Spearman's ρ; iii) ranking in concordance order and iv) a central limit theorem that characterizes stochastic behavior of Monte Carlo estimators when the sample size tends to infinity. The proposed simulation framework inherits the simplicity of the standard antithetic sampling method, requiring the definition of a set of reference points in the sampling space and the generation of uniform numbers on the segments joining the points. We provide applications to Monte Carlo integration and Markov Chain Monte Carlo Bayesian estimation.

C0504: **One integral transform of the copula function**
Presenter: **Bozidar Popovic**, University of Montenegro, Montenegro

An integral transformation of the copula function is studied. Under certain conditions, the integral transformation of a given copula function is also the copula function. Some properties of the integral transformation of a copula function are studied. Also, the bounds of Kendall's tau and Spearman's rho are derived.

CO140 Room Aula C COMPUTATIONAL STATISTICS: THEORY AND APPLICATIONS Chair: Alba Martinez-Ruiz

C0308: **Sampling redesign considering spatial t-Student models: An effective sample size application**
Presenter: **Luciana Pagliosa Carvalho Guedes**, Universidade Estadual do Oeste do Parana, Brazil
Co-authors: Leticia Ellen Dal Canton, Miguel Angel Uribe-Opazo, Tamara Cantu Maltauro, Rosangela Aparecida Botinha Assumpcao

The spatial dependence modeling of georeferenced variables, in the presence of outliers, should consider the use of a robust probability distribution, such as the Student's t-distribution. This distribution can reduce the influence of outliers in the spatial dependence structure. Also, the financial investment in data collection and laboratory analysis of soil samples is a relevant factor when mapping agricultural areas. In this context, the reduction of sample size can be performed by the univariate effective sample size methodology (ESS), assuming that the t-Student model represents the probability distribution. An application of the ESS to redesign a sample configuration in an agricultural area with 167.35 hectares cultivated with soybean is presented to analyze the spatial dependence of soil penetration resistance with outliers.

C0404: **A distributed algorithm for exhaustive normality test**
Presenter: **Ruben Carvajal-Schiaffino**, DMCC - Universidad de Santiago de Chile, Chile

The normal distribution assumption of data is necessary for a wide variety of statistical analyses, which are valid once normality tests are passed. Regarding the multivariate normality, algorithms have been developed to test it. However, they are often -near always- applied only to complete samples, excluding subsets of variables. As this number increases, an exhaustive analysis becomes less practical and also a deviation of normality within subspaces in the sample becomes more probable. Most statistical packages run in only one process, without the direct possibility of running distributed algorithms, which could allow a considerable time-saving. A distributed algorithm for running distributed normality tests is introduced. In contrast with the sequential solution, it means a viable solution for this problem.

C0307: **Incremental SVD for some numerical aspects of multiblock redundancy analysis and big data streams**
Presenter: **Alba Martinez-Ruiz**, Universidad Diego Portales, Chile
Co-authors: Carlo Lauro

Based on the incremental SVD algorithm, a new procedure is proposed to solve the decomposition problem found by multiblock redundancy analysis when analyzing streaming data, i.e. data that is generated continuously. The redundancy procedure involves the SVD of a square and symmetric matrix of $q \times q$, where q is the number of variables in the endogenous block of variables. If q is large, the factorization is a time- and resource-consuming task, much more if the matrix is continuously updated in real-time. A good strategy is analyzing the data in small sets that are continually updated. To preserve the column-wise formulation of the incremental SVD algorithm, we derived the column-wise variant of the redundancy method and implemented an incremental approach for the procedure. Numerical experiments are reported to illustrate the accuracy and performance of the incremental solution for analyzing streaming multiblock data. In addition, we report results for examining how the incremental SVD algorithm approximates singular vectors when varying a forgetting factor and the number of significant singular vectors kept at each iteration. The results provide evidence about the suitability of the new approach for the analysis of large streaming data.

C0400: **Assessing the estimation of nearly singular covariance matrices for modeling spatial variables**
Presenter: **Jonathan Acosta**, Pontificia Universidad Catolica de Chile, Chile
Co-authors: Ronny Vallejos

Spatial analysis commonly relies on the estimation of a covariance matrix associated with a random field. This estimation strongly impacts the variogram estimation and prediction where the process has not been observed, which in turn influences the construction of more sophisticated models. If some of the distances between all the possible pairs of observations in the plane are small, then we may have an ill-conditioned problem that results in a nearly singular covariance matrix. We suggest a covariance matrix estimation method that works well even when there are very close pairs of locations on the plane. Our method is an extension to a spatial case of a method that is based on the estimation of eigenvalues of the unitary matrix decomposition of the covariance matrix. Several numerical examples are conducted to provide evidence of a good performance of the suggested method; especially in estimating the variance components of a spatial regression process. In addition, an application to macroalgae estimation in a restricted area in the Pacific Ocean is developed to determine a suitable estimation of the effective sample size associated with the transect sampling scheme.

C0402: **Regularised PCA for incremental single imputation of missings**
Presenter: **Alfonso Iodice D Enza**, University of Naples Federico II, Italy
Co-authors: Angelos Markos, Francesco Palumbo

Principal Component Analysis (PCA) is an eigendecomposition of a properly transformed matrix, then its standard application requires the data set to be complete (no missing entries). Alternative implementations have been proposed in the literature that extends the PCA to incomplete data sets. Recent comparative reviews of PCA algorithms with missings proved regularised iterative PCA algorithm (RPCA) to be effective. In some applications, incomplete data are constantly produced (e.g. process sensor data) and the corresponding data flow is often analysed in chunks (subsets of observations). In this setting, RPCA could be applied to each chunk, with the result that the PCA solutions (and, the imputations) of single chunks are independent from one another. An incremental RPCA implementation is proposed such that the imputation of each new chunk is based on that chunk, and on all the chunks analysed that far. The proposed procedure is compared to batch RPCA considering different data sets and missing data mechanisms. Experimental results show that the incremental approach has an appreciable performance when the data is missing not completely at random, and the first analysed chunks contain sufficient information on the data structure.

CO049 Room Aula H OPTIMAL EXPERIMENTAL DESIGN AND APPLICATIONS **Chair: Ellinor Fackle-Fornius**

C0187: **A design criterion for symmetric model discrimination based on flexible nominal sets**
Presenter: **Werner Mueller**, Johannes Kepler University Linz, Austria
Co-authors: Radoslav Harman

Experimental design applications for discriminating between models have been hampered by the assumption of knowing beforehand which model is the true one, which is counter to the very aim of the experiment. Previous approaches to alleviate this requirement were either symmetrizations of asymmetric techniques, or Bayesian, minimax, and sequential methods. We present a genuinely symmetric criterion based on a linearized distance between mean value surfaces and the newly introduced tool of flexible nominal sets. We demonstrate the computational efficiency of the approach using the proposed criterion and provide a Monte Carlo evaluation of its discrimination performance based on the likelihood ratio. An application for a pair of competing models in enzyme kinetics is given.

C0291: **Optimal designs for comparing regression curves: Dependence within and between groups**
Presenter: **Kirsten Schorning**, Technical University Dortmund, Germany
Co-authors: Holger Dette

The focus is on the problem of designing experiments for the comparison of two regression curves describing the relation between a predictor and a response in two groups, where the data between and within the group may be dependent. In order to derive efficient designs, we use results from stochastic analysis to identify the best linear unbiased estimator (BLUE) in a corresponding continuous model. It will be demonstrated that in general simultaneous estimation using the data from both groups yields more precise results than estimation of the parameters separately in the two groups. Using the BLUE from simultaneous estimation, we then construct an efficient linear estimator for finite sample size by minimizing the mean squared error between the optimal solution in the continuous model and its discrete approximation with respect to the weights (of the linear estimator). Finally, the optimal design points are determined by minimizing the maximal width of a simultaneous confidence band for the difference of the two regression functions. The advantages of the new approach are illustrated by means of a simulation study, where it is shown that the use of the optimal designs yields substantially narrower confidence bands than the application of uniform designs.

C0373: **Computing optimal designs of multifactor experiments**
Presenter: **Lenka Filova**, Comenius University in Bratislava, Slovakia
Co-authors: Radoslav Harman, Samuel Rosa

A method is described for computing efficient approximate experimental designs for models with multiple independent discrete factors. The algorithm that we propose can be used to solve problems with an immense number of combinations of factor levels, e.g., 5 factors, each with 1000 levels, and obtain an optimal design in several seconds. The proposed algorithm alternates between two key steps: 1) the construction of exploration sets composed of star-shaped components and separate, highly informative design points and 2) the application of a conventional method for computing optimal approximate designs on medium-sized design spaces. We illustrate the performance of the algorithm on several nonlinear statistical models used in practice.

C0495: **Optimal pretesting of questions for Swedish national tests in school**
Presenter: **Frank Miller**, Stockholm University, Sweden
Co-authors: Ellinor Fackle-Fornius

Questions for future national tests are usually pretested by voluntary pupils who already have participated in their ordinary national test. Instead of allocating the pretest questions randomly to different classes, the 2022 pretesting for mathematics allocates the questions based on the pupils results from their ordinary national test, i.e., based on the ability of each pupil. We will describe the optimal experimental design method which we used to allocate the questions to the pupils in this years pretesting. Item response theory models were used for each question to describe the probability to answer correctly or to receive a certain number of points. The difficulty of a question and other characteristics are supposed to be estimated as precise as possible in the pretesting. Using a simulated annealing algorithm, we determined a D-optimal design for question-allocation and incorporated some specific constraints for this problem. We show that the chosen design is considerably more efficient than the random allocation of questions.

C0511: **Optimal dose-finding for drug combinations**
Presenter: **Renata Eirini Tsirpitzi**, Stockholm University, Sweden
Co-authors: Frank Miller

A combination of drugs is frequently done in clinical praxis. Therefore, the effect of a combination treatment should be investigated in clinical studies before. We explore the optimal combination and the optimal doses by improving the efficacy when two drugs are simultaneously present. We consider an efficacy Emax model for combining two drugs by including an interaction parameter. There are three possible interactions between the two drugs. When the interaction parameter is equal to 0, then there is no interaction and is called additivity. If the parameter is positive, then we have synergy between the two drugs. When the deviation of the interaction term from 0 is negative, then the interaction is considered antagonism. The optimality criterion used is the D-optimality and the results were obtained by applying our own developed algorithm which is based on the Fedorov algorithm. The results indicate that the number of doses, the doses, and their weights depend strongly on the model parameters.

CO183 Room Aula E STOCHASTIC MODELS FOR DYNAMICAL SYSTEMS: METHODS AND COMPUTATIONS Chair: Manuel Molina

C0251: **Coalescence in branching processes with age dependent structure in population**
Presenter: **Sumit Kumar Yadav**, Indian Institute of Technology Roorkee, India
Co-authors: Prof Arnab K Laha

Branching processes and their variants are widely used mathematical models in studying population dynamics. In the recent past, branching processes have also found applications in areas like operations research, marketing, finance, genetics etc. A problem that has caught attention in the context of coalescence in branching processes is as follows: Assume that one individual starts the branching process in 0-th generation and the population size of the tree obtained by the branching process in generation n is greater than 1. Next, pick two individuals from n-th generation at random and trace their lines of descent back till they meet. Call that random generation by $X(n)$. The objective is to study the properties of $X(n)$. While this problem has been studied by many authors for simple and multitype discrete time branching processes, not much attention has been given to the realistic extension when one individual is allowed to survive for more than one generation and can also give birth more than once. We study this problem for some deterministic and random cases. Explicit expressions about some mathematical properties of $X(n)$ have been derived for broad classes of deterministic trees. For random trees, we provide an explicit expression for some special cases. We also derive properties of $X(n)$ as n goes to infinity. A simulation analysis has also been performed, and some interesting insights are discussed.

C0364: **From multi-type age structure models to epidemic compartmental models**
Presenter: **Jie Yen Fan**, University of Sydney, Australia

Often, there is a heterogeneous system to model, such as a population where its dynamic depends on the age and other characteristics of the individuals. A general multi-type age structure population model, along with its law of large numbers and central limit theorem, will be discussed. Some examples, including sexual reproduction and the spread of viral infection, will be given.

C0368: **Subcritical multitype Markov branching processes with immigration generated by Poisson random measures**
Presenter: **Maroussia Slavtchova-Bojkova**, Sofia University, Bulgaria
Co-authors: Ollivier Hyrien, Nikolay Yanev

Multitype subcritical Markov branching processes with immigration driven by Poisson random measures are investigated. Limiting distributions are established for various rates of the Poisson measures when they are asymptotically equivalent to exponential or regularly varying functions. Results analogous to a strong LLN are proved, and limiting normal distributions are obtained when the local intensity of the Poisson measure increases with time. When it decreases, conditional limiting distributions are established. A stationary limiting distribution is obtained when the growth rate of the Poisson mean measure is asymptotically linear. The asymptotic behavior of the first and second moments of the processes is investigated as well.

C0317: **Fundamental problems arising in the analysis of applied stochastic models**
Presenter: **Elena Yarovaya**, Lomonosov Moscow State University, Russia

The aim is to study the behavior of some stochastic processes describing the evolution of systems with a complex structure. The main focus is on models involving the generation and transport of particles, so-called branching random walks. In recent years, branching random walks have become a rapidly developing area of stochastic processes. Special attention is paid to the analysis of the asymptotic behavior of particle numbers and their moments for symmetric BRWs with a few sources of branching and a finite or infinite number of initial particles under various assumptions on the variance of random walk jumps. The proofs of some limit theorems for BRWs with a finite number of sources and pseudo-sources, where violation of random walk symmetry is allowed, are often based on the verification of the Carleman condition, which guarantees the uniqueness of the definition of the limit probability distribution of particle numbers by their moments. In this context, questions about the relationship between such sufficient conditions based on the growth rate of the limiting moments of the particle numbers are discussed. For BRWs with branching sources at each point of the lattice, where the law of reproduction and death of particles is described by a critical branching process, limit theorems for the behavior of populations and subpopulations of particles are given. The research was supported by RFBR, project No. 20-01-00487.

C0217: **Stochastic modeling in dynamical populations through two-sex branching processes: Inferential and computational results**
Presenter: **Manuel Molina**, University of Extremadura, Spain
Co-authors: Manuel Mota, Alfonso Ramos

In the general context of stochastic modeling, branching processes are appropriate mathematical models to describe the probabilistic evolution of dynamical systems. They are an active research area of theoretical and practical interest with applicability to such fields as biology, demography, epidemiology, genetics, population dynamics, and others. Branching processes have especially played a major role in modeling the demographic dynamics of biological populations whose size evolves over time due to random births and deaths. In particular, in order to describe the dynamics of biological populations with sexual reproduction, several classes of two-sex branching processes have been introduced. We will focus the attention on the class of two-sex branching processes where several mating strategies and a variety of reproductive possibilities are considered. It is also assumed the most realistic situation which both phases, mating and reproduction, could be influenced by the numbers of females and males in the population. By considering the most general non-parametric statistical setting, several inferential and computational questions about the most informative reproductive parameters included in the mathematical model are investigated. As an illustration, the proposed methodology is applied to describe the demographic dynamics of some salmonid populations.

CC161 Room Aula D STATISTICAL MODELLING Chair: Shubhadeep Chakraborty

C0195: **Quantile LASSO with change-points in panel data models**
Presenter: **Matus Maciak**, Charles University, Czech Republic

Panel data are commonly used in all kinds of econometric problems under various regularity assumptions. We investigate the panel data models with changepoints and the atomic pursuit technique and quantile estimation are applied to obtain the final estimate. Robust estimates and a complex insight into the data generating mechanism are both achieved by adopting the quantile LASSO approach. The final model is produced in a fully data-driven manner in just one single step. The final estimate is, under some reasonable assumptions, shown to be consistent with respect to the model estimation and the changepoint detection performance. The finite sample properties are investigated in a simulation study and the proposed methodology is applied for the Apple call option pricing problem.

C0306: **A general joint latent class model of longitudinal and survival data with time-varying membership probability**
Presenter: **Ruoyu Miao**, The University of Manchester, United Kingdom

Joint latent class modelling has been developed considerably in the past two decades. In some instances, the models are linked by the latent class k (i.e. the number of subgroups), in others, they are joined by shared random effects or a heterogeneous random covariance matrix. We propose an extension to the joint latent class model (JLCM) in which probabilities of subjects being in latent class k can be set to vary with time. This can be a more flexible way to analyse the effect of treatments on patients. For example, a patient may be in the period I at the first visit time and may move to period II at the second visit time, implying the treatment the patient had before might be noneffective at the following visit time. For a dataset with these particular features, the joint latent class model which allows jumps among different subgroups can potentially provide more information as well as more accurate estimation and prediction results compared to the basic JLCM. A Bayesian approach is used to do the estimation and a

DIC criterion is used to decide the optimal number of classes. Simulation results indicate that the proposed model produces accurate results and the time-varying JLCM outperforms the basic JLCM. We also illustrate the performance of our proposed JLCM on the aids data.

C0433: **Uncovering regions of maximum dissimilarityon random process data**

Presenter: **Gabriel Martos**, Fundacion Universidad Torcuato Di Tella, Argentina

Co-authors: Miguel de Carvalho

The comparison of local characteristics of two random processes can shed light on periods of time or space at which the processes differ the most. A method is proposed that learns about regions with a certain volume, where the marginal attributes of two processes are less similar. The proposed methods are devised in full generality for the setting where the data of interest are themselves stochastic processes, and thus the proposed method can be used for pointing out the regions of maximum dissimilarity with a certain volume, in the contexts of functional data, time series, and point processes. The parameter functions underlying both stochastic processes of interest are modeled via a basis representation, and Bayesian inference is conducted via an integrated nested Laplace approximation. The numerical studies validate the proposed methods, and we showcase their application with case studies on criminology, finance, and medicine.

C0674: **Prior weights of Dirichlet PDFs**

Presenter: **Audun Josang**, University of Oslo, Norway

In the model to be proposed, a Dirichlet PDF has an uninformative prior weight which is initially equal to the domain cardinality k, and decreases to become equal to a convergence constant C as the amount of observation evidence increases. The advantage of this approach is that the vacuous Dirichlet PDF (i.e. in the absence of evidence) is always uniform, while at the same time ensuring that the prior carries low weight relative to the observation evidence irrespective of the domain cardinality. More formally, the evidence of a Dirichlet PDF over a multidimensional domain X is denoted as a vector alpha expressed as: $\alpha(x) = r(x) + a(x)W$, where $r(x) \geq 0$ for all $x \in X$. The prior probability distribution is denoted by the vector a, and the observation evidence is represented by the vector r. With these parameters defined, the uninformative prior weight denoted by W can be expressed as: $W = (k + Ck\sum[r(x)])/(1 + k\sum[r(x)])$. The convergence uninformative prior weight C determines the sensitivity of the Dirichlet PDF to new observation evidence. The larger C, the less sensitive the Dirichlet PDF becomes to new observation evidence. If we assume that the sensitivity should always be the same irrespective of the domain cardinality, then it is natural to set $C = 2$, which reflects the same sensitivity as for the uninformative prior weight of the Beta PDF over a binary domain.

C0281: **Weighted average least squares for negative binomial regression**

Presenter: **Kevin Huynh**, University of Basel, Switzerland

Model averaging methods have become an increasingly popular tool for improving predictions and dealing with model uncertainty, especially in Bayesian settings. Recently, frequentist model averaging methods, such as information theoretic and least squares model averaging, have emerged. The focus is on the issue of covariate uncertainty, where managing the computational resources is key: The model space grows exponentially with the number of covariates such that averaged models must often be approximated. Weighted average least squares (WALS), first introduced for (generalized) linear models in the econometric literature, combines Bayesian and frequentist aspects and additionally employs a semiorthogonal transformation of the regressors to reduce the computational burden. WALS is extended for generalized linear models to the negative binomial (NB) regression model for overdispersed count data. The predictive power of WALS for NB regression is compared to traditional estimators in a simulation experiment and in an empirical application using data on doctor visits.

CC211 Room Aula I MIXTURE MODELS — Chair: Christian Hennig

C0276: **Structured Dirichlet mixtures as priors for generalised entropy estimation**

Presenter: **Tanita Botha**, University of Pretoria, South Africa

Co-authors: Johan Ferreira, Andriette Bekker

Entropy indicates an amount of information contained in a system, and the suitable estimation of entropy continues to receive ongoing focus particularly in the case of multivariate data. Data on the unit simplex are often found in different spheres in science; particularly arising from compositional data. In these instances, the Dirichlet distribution is frequently employed as model of choice as prior in a Bayesian approach, but theoretically only accounts for possibilities of negatively correlated proportions. The purpose is to implement previously unconsidered mixtures of the Dirichlet distribution as a prior for the multinomial model that hypothetically allows for positive correlated data. Some statistical properties are briefly derived and the derived and fitted posterior model is used to obtain insight into the behaviour of some entropy forms for potential prior selection using real data, and a prior selection method is implemented to suggest a suitable prior for the consideration of the practitioner.

C0389: **Component and feature selection in mixtures of generalised linear models**

Presenter: **Sollie Millard**, University of Pretoria, South Africa

Co-authors: Salomi Millard, Frans Kanfer, Mohammad Arashi, Gaonyalelwe Maribe

Datasets with a relatively large number of highly correlated features are often found in applications of a finite mixture of regression models, resulting in unstable parameter estimates. The contribution of each feature towards the response variable differs in the respective components of the mixture model. This creates a complex feature selection problem. Penalised regression methods are frequently used to perform feature selection whilst addressing the issues that arise due to multicollinearity. The estimation of a mixture of generalised linear models in the presence of multicollinearity is considered, addressing both feature and component selection. The selection of the optimal number of components is important since traditional maximum likelihood estimation faces difficulty when an incorrect number of components are specified. We propose a novel penalised-likelihood approach to conduct model selection for finite mixtures of generalised linear models. Penalties are imposed on both mixing proportions and regression coefficients, hence order selection of the mixture and the variable selection in each component can be simultaneously achieved. A modified EM algorithm is proposed. We consider the use of the novel modified elastic-net penalty for feature selection. An extensive simulation study is performed to demonstrate the properties pertaining to component and feature selection of this approach.

C0595: **Evidence estimation in finite and infinite mixture models and applications**

Presenter: **Adrien Hairault**, Universite Paris Dauphine PSL, France

Co-authors: Christian Robert, Judith Rousseau

Estimating the model evidence - or marginal likelihood of the data - is a notoriously difficult task for finite and infinite mixture models and we reexamine here different Monte Carlo techniques advocated in the recent literature, as well as novel approaches based on the reverse logistic regression technique, Chib's algorithm, and Sequential Monte Carlo (SMC). Applications are numerous. In particular, testing for the number of components in a finite mixture model or against the fit of a finite mixture model for a given dataset has long been and still is an issue of much interest, albeit yet missing a fully satisfactory resolution. Using a Bayes factor to find the right number of components K in a finite mixture model is known to provide a consistent procedure. We furthermore establish the consistency of the Bayes factor when comparing a parametric family of finite mixtures against the nonparametric "strongly identifiable" Dirichlet Process Mixture (DPM) model.

C0655: **Identification of high-energy astrophysical point sources via hierarchical Bayesian nonparametric clustering**

Presenter: **Andrea Sottosanti**, University of Padua, Italy

Co-authors: Mauro Bernardi, Alessandra Rosalba Brazzale, Alex Geringer-Sameth, David Stenning, Roberto Trotta, David van Dyk

The light we receive from distant astrophysical objects carries information about their origins and the physical mechanisms that power them.

The study of these signals, however, is complicated by the fact that observations are often a mixture of the light emitted by multiple localized sources situated in a spatially-varying background. A general algorithm to achieve robust and accurate source identification remains an open question in astrophysics. The focus is on high-energy light (such as X-rays and gamma-rays), for which observatories can detect individual photons (quanta of light), measuring their incoming direction, arrival time, and energy. The proposed Bayesian methodology uses both the spatial and energy information to identify point sources, that is, separate them from the spatially-varying background, estimate their number, and compute the posterior probabilities that each photon originated from each identified source. This is accomplished via a Dirichlet process mixture while the background is simultaneously reconstructed via a flexible Bayesian nonparametric model based on B-splines. Our proposed method is validated with a suite of simulation studies and illustrated with an application to a complex region of the sky observed by the Fermi Gamma-ray Space Telescope.

C0432: **Estimation of parameters of a mixture of two exponential distributions**
Presenter: **Trijya Singh**, Le Moyne College, Syracuse, NY, United States

For estimating the parameters of a mixture of two exponential distributions, the method of moments, which uses roots of a quadratic equation involving the estimates of the first three raw moments has been used in the past. Because of poor estimates of these moments, in many situations, the roots of the quadratic equation turn out to be complex and hence the method fails. A methodology based on a quadrature formula of numerical integration is proposed for the estimation of the moments. These moment estimates always produce real roots of the quadratic equation in the case of sampling from a mixture of two exponential distributions and produce estimates of the parameters. To fully capture the long tail or heavy tail behavior of mixture models, the peak and tail characteristics of a distribution that are explained by the standardized fourth central moment (the coefficient of kurtosis) are incorporated into the proposed methodology by using the first four sample moments. The estimates obtained by the method of moments are proposed as initial estimates for an optimization algorithm to obtain least squares estimates. Some important applications have been discussed using a drug concentration dataset, and it has been shown that methods using all four moments perform better than those based on only the first three moments.

CC155 Room Aula Q SEMI- AND NONPARAMETRIC METHODS — Chair: Stefan Sperlich

C0331: **Nonparametric comparison of epidemic time trends: The case of COVID-19**
Presenter: **Marina Khismatullina**, Erasmus University Rotterdam, Netherlands
Co-authors: Michael Vogt

The COVID-19 pandemic has been one of the most pressing issues for the past two years. A question which was particularly important for governments and policymakers is the following: Does the virus spread in the same way in different countries? Or are there significant differences in the development of the epidemic? We devise a new inference method for detecting differences in the development of the epidemic time trends across countries. Specifically, it allows making simultaneous confidence statements about the regions where the trends differ. In the theoretical part, we prove that the method controls the familywise error rate, that is, the probability of wrongly rejecting at least one null hypothesis. In our empirical study, we use the method to compare the outbreak patterns of the epidemic in a number of European countries.

C0442: **A refined Weissman estimator for extreme quantiles**
Presenter: **Stephane Girard**, Inria, France
Co-authors: Michael Allouche, Jonathan El Methni

Weissman's extrapolation methodology for estimating extreme quantiles from heavy-tailed distributions is based on two estimators: an order statistic to estimate an intermediate quantile and an estimator of the tail-index. The common practice is to select the same intermediate sequence for both estimators. We show how an adapted choice of two different intermediate sequences leads to a reduction of the asymptotic bias associated with the resulting refined Weissman estimator. The asymptotic normality of the latter estimator is established and a data-driven method is introduced for the practical selection of the intermediate sequences. Our approach is compared to Weissman estimator and to six bias-reduced estimators of extreme quantiles in a large-scale simulation study. It appears that the refined Weissman estimator outperforms its competitors in a wide variety of situations, especially in challenging high-bias cases. Finally, an illustration of an actuarial real data set is provided.

C0491: **Multiscale splines and local polynomials**
Presenter: **Maarten Jansen**, Universita libre de Bruxelles, Belgium

The benefits of non-linear estimation of piecewise smooth data through a sparse multiresolution decomposition, such as offered by a wavelet transform, can be incorporated into well-established methods, such as splines and kernel-based approaches, by upgrading the unit scale approach into a multiscale version, thus adding locality in scale (or frequency) to the already existing spatial locality. The combination of splines or kernel-based methods with a multiresolution analysis extends the scope of the latter beyond the dyadic, equispaced setting of most wavelet methods. The construction of the multiscale spline, kernel or local polynomial methods proceeds through the so-called lifting scheme. The understanding of this scheme is crucial for successful applications in nonparametric statistical estimation. In particular, two issues play an important role. On one hand, the smoothness of the reconstruction through the process of multiscale refinement should be monitored. On the other hand, the variance propagation throughout the scheme should be controlled, by looking at the singular value decomposition of the underlying projection matrix.

C0635: **Post-selection inference for partially linear high-dimensional single-index models**
Presenter: **Pieter Willems**, KU Leuven, Belgium
Co-authors: Gerda Claeskens

A post-model selection estimator and method for inference are introduced for the linear part of a partially linear model $Y = X\alpha + g(Z\gamma) + \varepsilon$. Typically the linear part consists of a fixed and limited number of variables $X = (X_1, \dots, X_r)$, while the control variables $Z = (Z_1, \dots, Z_p)$ might consist of a large number of variables, p, that is allowed to grow with the sample size n and potentially exceeds n. The function $g(\cdot)$ is not specified beforehand and is estimated via B-splines in combination with a l_1 regularization. This research is relevant because there is clearly an interest in developing new procedures that provide inference which is valid after model selection, which here takes place via the l_1 regularizer. This methodology allows for imperfect variable sections and provides a confidence region that is uniformly valid under certain assumptions. Inferential properties are established for this method by proving that asymptotic multivariate normality holds for the newly introduced estimator in the context of a partially linear single-index model. Simulation studies were conducted in multiple settings and a comparison was made with the methodology introduced by other authors in order to illustrate the empirical properties of the newly introduced estimator.

C0631: **Challenges in assessing lack of fit for non-parametric quantile models**
Presenter: **Ting Zhang**, McGill University, Canada
Co-authors: Gael Varoquaux, Jean-Baptiste Poline, Celia Greenwood

Assessing model fit in non-parametric quantile regressions with multiple predictors is challenging. A new paradigm has been proposed for testing lack-of-fit, by testing the equality of the two covariate distributions defined by separating the data at the fitted quantile. However, their test has limitations. (1) It detects underfit (e.g. a missing covariate) but not data overfit. (2) It uses data twice for model fitting and lack-of-fit testing, thereby leading to invalid type 1 errors. We propose to improve this testing procedure by: (1) splitting data into training and testing sets, (2) replacing the core test statistic for testing distributional equivalence by an L_1 kernel mean embedding, and (3) modifying the estimation of significance by changing the wild bootstrap method. We will first illustrate the problems through extensive simulation studies, and compare the proposed modifications (2) and (3) to the original lack-of-fit test after data splitting. Performance is assessed by type 1 error control, power and computational speed. Our replacement test statistic has better discrimination, and a known distribution making computations much faster than the previous method. However, since the true data generating model is always unknown, the lack of fit tests which use models that fit observed data is intrinsically problematic.

CC207 Room Aula F SPATIAL STATISTICS **Chair: Pier Giovanni Bissiri**

C0248: **Data fusion in a two-stage spatio-temporal model using the INLA-SPDE approach**

Presenter: **Stephen Jun Villejo**, University of Glasgow, United Kingdom

Co-authors: Janine Illian, Ben Swallow

A two-stage model is proposed, motivated by an epidemiological problem which involves data with different spatial supports. The response is areal, while the predictor data are measurements from a geostatistical process and high-resolution outputs either from satellites or numerical models. The first stage assumes a common latent field for the geostatistical and the high-resolution data, whereby both are error-prone realizations of the field. The spatial effect of the latent field is assumed to evolve in time, inducing spatiotemporal dependence and is modelled using a stochastic partial differential equation approach. This provides a Markov structure on the random field, speeding up computation and spatial interpolation. The second stage fits a GLMM using spatial averages of the estimated latent field, and additional spatial and temporal random effects. The latent Gaussian models are estimated using the integrated nested Laplace approximation, a deterministic Bayesian inference approach. Uncertainty from the first stage is accounted for by simulating several times from the posterior predictive distribution of the latent field. A simulation study was done to assess the impact of the sparsity of the data, length of time, and priors specification on the model fit. The method was also applied to actual data in England.

C0524: **Spatial meshing and manifold preconditioning for Bayesian analysis of non-Gaussian data**

Presenter: **Michele Peruzzi**, Duke University, United States

Co-authors: David Dunson

Quantifying spatial associations in multivariate geolocated data of different types is achievable via spatial random effects in a Bayesian hierarchical model, but severe computational bottlenecks arise when spatial dependence is encoded as a latent Gaussian process (GP) in the increasingly common large-scale data settings on which we focus. The scenario worsens in non-Gaussian models because the reduced analytical tractability leads to additional hurdles to computational efficiency. We introduce methodologies for efficiently computing multivariate Bayesian models of spatially referenced non-Gaussian data. First, we outline spatial meshing as a tool for building scalable processes using patterned directed acyclic graphs. Then, we introduce a novel Langevin method which achieves superior sampling performance with non-Gaussian multivariate data that are common in studying species' communities. We proceed with outlining strategies for improving Markov-chain Monte Carlo performance in the settings on which we focus. We conclude with extensions and applications showcasing the flexibility of the proposed methodologies and the publicly-available software package.

C0601: **Bayesian variable selection in double generalized linear Tweedie spatial process models**

Presenter: **Aritra Halder**, University of Virginia, United States

Double generalized linear models provide a flexible framework for modeling data by allowing the mean and the dispersion to vary across observations. Common members of the exponential dispersion family including Gaussian, compound Poisson-gamma, Gamma, and inverse-Gaussian, are known to admit such models. However, the lack of their use can be attributed to ambiguities that exist in the model specification under a large number of covariates and complications that arise when data from a chosen application displays dependence. We consider a hierarchical specification for these models with a spatial random effect. The spatial effect is targeted at performing uncertainty quantification by modeling dependence within the data arising from location-based indexing of the response. We focus on a Gaussian process specification for the spatial effect. Simultaneously we tackle the problem of the model specification under such hierarchical spatial process models using Bayesian variable selection, which is effected through a continuous spike and slab prior on the model parameters (or fixed effects). The novelty lies in the Bayesian frameworks developed for such models which have not been explored previously. We perform various synthetic experiments to showcase the accuracy of our frameworks. These developed frameworks are then applied to analyse automobile insurance premiums in Connecticut.

C0660: **A conditional Gaussian process model for ordinal data and its application in predicting herbicidal performance**

Presenter: **Arron Gosnell**, University of Bath, United Kingdom

With the proliferation of screening tools for chemical testing, it is now possible to create vast databases of chemicals easily. On the other hand, the development of a rigorous statistical methodology that can be used to analyse these large databases is in its infancy, and further development to facilitate chemical discovery is imperative. Current methods employed to analyse these data fail to incorporate the chemical structure of the tested compound, and as a result, this feature is unaccounted for in the model. We will discuss the Tanimoto similarity as a measure of closeness between chemical compounds and its use within a Gaussian process model. We will demonstrate the application of the proposed model for analysing data from agricultural experiments to assess the herbicidal performance of chemical compounds. The response variable is ordinal, so a proportional odds model is used, with the cumulative probabilities being functions of the Gaussian process. We will show that accounting for correlation results in improved model performance over a simple mixed-effects model and an alternative random forests model. We will discuss the tools used to overcome certain hurdles in developing the model and the use of proper scoring rules to evaluate model performance.

C0574: **Latent Gaussian model boosting**

Presenter: **Fabio Sigrist**, Lucerne University of Applied Sciences, Switzerland

Latent Gaussian models and boosting are widely used techniques in statistics and machine learning. Latent Gaussian models, such as Gaussian process and grouped random-effects models, are flexible prior models that allow for making probabilistic predictions. However, existing latent Gaussian models usually assume either a zero or a linear prior mean function which can be an unrealistic assumption. Tree-boosting shows excellent predictive accuracy on many data sets, but potential drawbacks are that it assumes conditional independence of samples, produces discontinuous predictions for, e.g., spatial data, and it can have difficulty with high-cardinality categorical variables. We introduce a novel approach that combines boosting and latent Gaussian models in order to remedy the above-mentioned drawbacks and leverage the advantages of both techniques. We obtain increased predictive accuracy compared to existing approaches in both simulated and real-world data experiments.

Thursday 25.08.2022 09:00 - 11:00 Parallel Session I – COMPSTAT2022

CV194 Room Virtual Room R1 TIME SERIES (VIRTUAL) Chair: Francesco Violante

C0416: **High-dimensional time series segmentation via factor-adjusted vector autoregressive modelling**
Presenter: **Hyeyoung Maeng**, Lancaster University, United Kingdom
Co-authors: Haeran Cho, Idris Eckley, Paul Fearnhead

Piecewise stationarity is a widely adopted assumption for modelling non-stationary time series. However, fitting piecewise stationary vector autoregressive (VAR) models to high-dimensional data is challenging as the number of parameters increases as a quadratic of the dimension. Recent approaches to address this have imposed sparsity assumptions on the parameters of the VAR model, but such assumptions have been shown to be inadequate when datasets exhibit strong (auto)correlations. We propose a piecewise stationary time series model that accounts for pervasive serial and cross-sectional correlations through a factor structure, and only assumes that any remaining idiosyncratic dependence between variables can be modelled by a sparse VAR model. We propose an accompanying two-stage change point detection methodology which fully addresses the challenges arising from not observing either the factors or the idiosyncratic VAR process directly. Its consistency in estimating both the total number and the locations of the change points in the latent components is established under conditions considerably more general than those in the existing literature. We demonstrate the competitive performance of the proposed methodology on simulated datasets and an application to US blue chip stocks data.

C0609: **Testing for the Sharpe ratio under a family of GARCH models**
Presenter: **Yifan Zhang**, Renmin University of China, China
Co-authors: Zhenya Liu, Shixuan Wang

The asymptotic properties of the Sharpe ratio estimator under a family of GARCH models are investigated. For financial time series, especially asset return, the stylized facts of left-skewed distribution and heteroscedasticity are often observed. In such circumstances, the limit behavior of the Sharpe ratio estimator is derived for the general GARCH(1,1) return. Additionally, we develop a strongly consistent estimator for the obtained asymptotic variance. A zero Sharpe ratio t-type test is proposed based on the theoretical results. Simulation studies demonstrate that the test has a good finite-sample performance under several specific examples. We illustrate the test in applications to explore the cross-sectional distribution of managerial skills measured with the Sharpe ratio in U.S. mutual funds.

C0630: **SHARP: A state-space HAR model with particle GIBBS sampling**
Presenter: **Aya Ghalayini**, Lancaster University, United Kingdom
Co-authors: Marwan Izzeldin, Mike Tsionas

The aim is to propose a general state-space autoregressive (AR) model with time-varying coefficients that follow an AR process with stochastic volatility. We implement these new specifications in the HAR framework to capture the time-varying salient feature of volatility using a two-state representation via a) allowing the time-varying coefficients to follow an AR(1) specification. b) introducing stochastic volatility for the innovations of the coefficients. Using high-frequency data of the SPY-ETF and representative NYSE stocks from 2000 to 2016, we show that the proposed model estimated using particle Gibbs sampling consistently outperforms different HAR model specifications in forecasting financial volatility.

C0264: **On time-dependent cointegration with an application**
Presenter: **Guy Melard**, Universita libre de Bruxelles, Belgium

The focus is on VARMA models with deterministically time-dependent (td), coefficients, possibly dependent on the series length n. Previous work studied the asymptotic theory of the quasi-maximum likelihood (QML) estimators for these tdVARMA$^{(n)}$) models. Under appropriate assumptions, these estimators are consistent and asymptotically normal. Time-dependent co-integration is considered. This is done by starting from a time-dependent extension of an error correction model (ECM), denoted tdECM$^{(n)}$. Generalizing previous work, the tdECM$^{(n)}$ is expressed as a tdVARMA$^{(n)}$ model on the differenced series. Hence, the original parameters of the tdECM$^{(n)}$ can be estimated by using a QML estimation method using the exact Gaussian likelihood. The asymptotic theory is applicable, and its assumptions can be checked a posteriori. An example on the US interest rates taken from the literature will illustrate the results: a time-dependent co-integration relation exists and is statistically significant.

C0378: **Asymptotic inference for a sign-double autoregressive (SDAR) model**
Presenter: **Emma Iglesias**, University of A Coruna (SPAIN), Spain

An extension of the double autoregressive (DAR) model is proposed: the sign-double autoregressive (SDAR) model, in the spirit of the GJR-GARCH model (also named the sign-ARCH model). Our model shares the important property of DAR models where a unit root does not imply nonstationarity and allows for asymmetry. We establish consistency and asymptotic normality of the quasi-maximum likelihood estimator in the context of the SDAR model. Furthermore, it is shown by simulations that the asymptotic properties also apply in finite samples. Finally, an empirical application shows the usefulness of our model.

CO168 Room Aula G TUTORIAL II Chair: Fabrizio Durante

C0613: **Tail dependence with copulas**
Presenter: **Fabrizio Durante**, University of Salento, Italy

Copula models have numerous advantages in describing the behavior of a multivariate stochastic system because of their flexibility in capturing various dependence aspects, especially in the tail of the joint distribution. We present several selected tools to describe the tail behaviour of random vectors by means of the copula approach. We start with a critical discussion about the classical tail dependence coefficients and their use. Then we consider some alternatives that are especially of interest in high dimensions. Computational aspects, as well as some applications, will be illustrated.

CO067 Room Aula B RECENT DEVELOPMENT IN THE NETWORK DATA ANALYSIS (VIRTUAL) Chair: Frederick Kin Hing Phoa

C0301: **Eliminating the biases of user influence and item popularity in bipartite networks**
Presenter: **Hohyun Jung**, Sungshin Women's University, Korea, South

User-item bipartite networks consist of users and items, where edges indicate the interactions of user-item pairs. We propose a Bayesian generative model for the user-item bipartite network that can measure the two types of rich-get-richer biases: item popularity and user influence biases. Furthermore, the model contains a novel measure of an item, namely the item quality that can be used in the item recommender system. The item quality represents the genuine worth of an item when the biases are removed. The Gibbs sampling algorithm alongside the adaptive rejection sampling is presented to obtain the posterior samples to perform the inference on the parameters. Monte Carlo simulations are performed to validate the presented algorithm. We apply the proposed model to Flickr user-tag and Netflix user-movie networks to yield remarkable interpretations of the rich-get-richer biases. We further discuss genuine item quality using Flickr tags and Netflix movies, considering the importance of bias elimination.

C0327: **Forecast reconciliation using linear models: Study on time series with network structure**
Presenter: **Mahsa Ashouri**, Academia Sinica, Taiwan
Co-authors: Sadid Sahami, Frederick Kin Hing Phoa

Forecasting hierarchical or grouped time series using a reconciliation approach involves two steps: computing base forecasts and reconciling the forecasts. Base forecasts can be computed by popular time series forecasting methods such as Exponential Smoothing (ETS) and Autoregressive Integrated Moving Average (ARIMA) models. The reconciliation step is a linear process that adjusts the base forecasts to ensure they are coherent. However, using ETS or ARIMA for base forecasts can be computationally challenging when there are a large number of series to forecast, as each model must be numerically optimized for each series. We propose a linear model that avoids this computational problem and handles the forecasting and reconciliation in a single step. This approach can also be extended to the network time series structure. This extended framework is used for forecasting if we have the network structure at each hierarchy level which applies the Least Absolute Shrinkage and Selection Operator (LASSO) to justify network connections. We illustrate our method using the export Free-on-Board (FOB) dataset.

C0363: **Quality of life and multilevel contact networks: Online study among healthy adults in Taiwan**
Presenter: **Tso-Jung Yen**, Academia Sinica, Taiwan

People's quality of life diverges on their demographics, socioeconomic status, and social connections. By taking both demographic and socioeconomic features into account, we investigated how the quality of life varied on social networks using data from both longitudinal surveys and contact diaries in a year-long (2015-2016) study. Our 4-wave, repeated measures of quality of life followed the brief version of the World Health Organization Quality of Life scale (WHOQOL-BREF). In our regression analysis, we integrated these survey measures with key time-varying and multilevel network indices based on contact diaries. People's quality of life may decrease if their daily contacts contain high proportions of weak ties. In addition, people tend to perceive a better quality of life when their daily contacts are face-to-face or initiated by others or when they contact someone who is in a good mood or someone with whom they can discuss important life issues. Our findings imply that both functional and structural aspects of the social network play important but different roles in shaping people's quality of life.

C0380: **Measuring uniqueness and diversity from a network perspective**
Presenter: **Wei-chung Liu**, Institute of Statistical Science, Academia Sinica, Taiwan, Taiwan

An index for quantifying node uniqueness in a network is developed. Our rationale is based on how effects from nodes can spread to all others, forming an interaction structure for the whole network. Using such an interaction structure, we then determine the interaction pattern of each node, and the similarity in the interaction pattern between a node and others then defines its uniqueness. A similar concept is then extended to compare the interaction patterns of all nodes, and an index is developed to measure the interaction diversity of a network. Our approach is of a very general nature, and we demonstrate it by analyzing 92 ecological network datasets. The relationship between our diversity measure and several network properties is then examined.

C0682: **Designing experiments for general network structures**
Presenter: **Ming-Chung Chang**, Institute of Statistical Science, Academia Sinica, Taiwan
Co-authors: Frederick Kin Hing Phoa, Jing-Wen Huang

Network experiments are commonly conducted in various fields, such as agriculture trials, medical experiments, and social networks. In these cases, an experimental unit may connect with some others, and the treatment applied to a unit has an effect on the responses of the neighboring units. Designing such experiments is rarely discussed in the literature. In this work, we study optimal network designs given unstructured treatments. Alphabetical optimality criteria are considered for selecting designs with high efficiency in estimating the treatment effects. We provide theoretical conditions for designs to be optimal and illustrate our theory with numerical examples.

CO129 **Room Aula E** **PIONEERING NEW FRONTIERS IN DISTRIBUTION AND MODELING** **Chair: Mohammad Arashi**

C0212: **Dirichlet distribution the superhero leading to robust innovations**
Presenter: **Andriette Bekker**, University of Pretoria, South Africa
Co-authors: Mohammad Arashi

The Dirichlet distribution is a well-known candidate for modeling compositional data sets. However, in the presence of outliers, this distribution fails to provide a robust model, adequate for outlying data sets, and these challenging issues require continuous exploration of alternative approaches. Such a drawback can be overcome by resorting to the beta-generating technique, which is a well-known mechanism in developing flexible models. The Kummer-Dirichlet distribution and the gamma distribution are coupled, and the development results in the proposal of the Kummer-Dirichlet gamma distribution, which has great flexibility in modeling. The method of maximum likelihood is applied in the estimation of the parameters. A model testing technique will be briefly reviewed to evaluate the performance. The usefulness of this newly proposed Dirichlet model is demonstrated through the application of synthetic and real data sets, where outliers are present. Finally, this candidate is briefly introduced as a model of prior under a Bayesian framework.

C0284: **A new family of multivariate centrally symmetric distributions**
Presenter: **Luca Bagnato**, Catholic University of the Sacred Heart, Italy
Co-authors: Antonio Punzo

A family of dimension-wise scaled normal mixtures (DSNMs) is proposed to model the joint distribution of a d-variate random variable with real-valued components. Each member of the family generalizes the multivariate normal (MN) distribution in two directions. Firstly, the DSNM has a more general type of symmetry with respect to the elliptical symmetry of the MN distribution and, secondly, the univariate marginals have similar heavy-tailed normal scale mixture distributions with (possibly) different tailedness parameters. As a consequence of practical interest, the DSNM allows for a different excess kurtosis on each dimension. We examine a number of properties of DSNMs and we describe two members of the DSNM family obtained in the case of components of the mixing random vector being either uniform or shifted exponential. These are examples of mixing distributions that guarantee a closed-form expression for the joint density of the DSNM. For the two DSNMs analyzed in detail, we describe algorithms, based on the expectation-maximization (EM) principle, to estimate the parameters by maximum likelihood. We use real data from the financial and biometrical fields to appreciate the advantages of our DSNMs over other symmetric heavy-tailed distributions available in the literature.

C0566: **Practical aspects of shape mixture constructions emanating from a Dirichlet setup**
Presenter: **JT Ferreira**, University of Pretoria, South Africa
Co-authors: Tanita Botha, Andriette Bekker

The Dirichlet distribution is arguably the most well-known multivariate distribution for implementation on the unitary simplex. Different generalisations exist, which include a shape mixture of Poisson weights known as the noncentral Dirichlet. This noncentral representation depends on the noncentrality parameters through the confluent hypergeometric function of several variables and admits both singly- and doubly noncentral representations; computational aspects are explored when the estimation of this singly and doubly noncentral Dirichlet is of interest. We investigate to what degree the additional parameter(s) and their effect on the doubly noncentral Dirichlet, compared to the singly alternative, affects the practical implementation of the model. Real data examples are used for this investigation by using maximum likelihood estimation for the parameters and further strengthened by simulation studies.

C0209: **A semi-parametric density estimation**
Presenter: **Mahdi Salehi**, University of Neyshabur, Iran
Co-authors: Andriette Bekker, Mohammad Arashi

A semi-parametric multivariate kernel density estimator is proposed with a more flexible family of kernels including skew-normal and skew-t. We show that the proposed estimator not only reduces boundary bias but also it is closer to the actual density compared to that of the usual estimator employing the Gaussian kernel. Finding optimum bandwidth under the mentioned asymmetric kernels is another main result where we shrink the bandwidth more than the one obtained under the normal assumption. Finally, through a numerical study, we will illustrate the application of the proposed semi-parametric kernel density estimator on density-based clustering using some simulated and real data sets.

C0303: **Robust modeling of multivariate heterogeneous datasets using a tractable multivariate skew heavy-tailed distribution**
Presenter: **Olcay Arslan**, Ankara University, Turkey
Co-authors: Fatma Zehra Dogru

Finite mixtures of multivariate normal distributions are often considered for modeling multivariate heterogeneous datasets. However, in applications, besides heterogeneity, datasets may have an asymmetric form with tail behavior different from the normal distribution so modeling them with a finite mixture of normal distributions may not provide an adequate model to represent all the features of the data. Therefore, recent research has focused on using finite mixtures of multivariate models with more flexible distributional forms to properly account for skewness, and heavy or light-tailedness to model multivariate heterogeneous data. As an alternative to the newly launched models in the literature, a finite mixture of multivariate skew Laplace normal (MSLN) distributions is introduced to simultaneously handle skewness and heavy tailedness in multivariate heterogeneous datasets. The MSLN distribution has recently been proposed with some plausible advantages over its counterparts. The proposed mixture model (FM-MSLN model) has some desirable properties, including a tractable density with a fewer number of parameters and ease of computation for simulation and estimation of parameters. Maximum likelihood parameter estimation of the FM-MSLN model via the expectation-maximization (EM) algorithm is given. The modeling performance of the FM-MSLN model is demonstrated using simulated datasets and a real data example.

C0376: **A copula-based measure of asymmetry between the lower and upper tail probabilities of bivariate distributions**
Presenter: **Shogo Kato**, The Institute of Statistical Mathematics, Japan
Co-authors: Toshinao Yoshiba, Shinto Eguchi

A measure of asymmetry between the lower and upper tail probabilities of bivariate distributions is proposed. The expression for the proposed measure can be simplified if bivariate distribution functions are represented using copulas. With this representation, it is seen that the proposed measure possesses some desirable properties as a measure of asymmetry. The limit of the proposed measure as the index goes to the boundary of its domain can be expressed in a simple form under certain conditions on copulas. A sample analogue of the proposed measure for a sample from a copula is presented and its weak convergence to a Gaussian process is shown. Another sample analogue of the presented measure is given, which is based on a sample from the original bivariate distribution on the plane. Simple methods for interval estimation are presented. As an example, the presented measure is applied to stock daily returns of S&P500 and Nikkei225.

CC150 Room Aula C CLUSTERING AND CLASSIFICATION Chair: Marta Nai Ruscone

C0461: **Combined-information criterion for clusterwise elastic-net regression: Application to omic data**
Presenter: **Stephanie Bougeard**, ANSES, France
Co-authors: Xavier Bry, Thomas Verron, Ndeye Niang

Many research questions pertain to a regression problem assuming that the population under study is not homogeneous with respect to the underlying model. In this setting, we propose an original method called Combined Information criterion CLUSterwise elastic-net regression (CICLUS). This method handles several methodological and application-related challenges. It is derived from both the information theory and the microeconomic utility theory and maximizes a well-defined criterion combining three weighted sub-criteria, each being related to a specific aim: getting a parsimonious partition, compact clusters for a better prediction of cluster-membership and a good within-cluster regression fit. The solving algorithm is monotonously convergent under mild assumptions. The CICLUS method provides an innovative solution to two key issues: the automatic optimization of the number of clusters and the issue of a prediction model. We applied it to elastic-net regression in order to be able to manage high-dimensional data involving redundant explanatory variables. CICLUS is illustrated through a real example in the field of omic data, showing how it improves the quality of the prediction and facilitates the interpretation. It should therefore prove useful whenever the data involve a population mixture such as, for example, in biology, social sciences, economics or marketing.

C0619: **Spatio-temporal clustering and classifying of seismic events in Chile**
Presenter: **Orietta Nicolis**, Universidad Andres Bello, Chile

Chile is one of the most seismic countries in the world, especially due to its particular position above a subduction zone where the Nazca tectonic plate dives down under the South America plate. This movement is responsible for a great number of seismic events, some of these with great magnitudes. Clustering seismic events allow for the characterization of the seismicity process besides defining the spatial region and the temporal window of the seismic events associated with a mainshock. The resulting clusters can be then used in classification models for identifying those events which may be precursors of great earthquakes. A new methodology for automatically clustering and classifying seismic events is proposed. First, a spatio-temporal DBSCAN (ST-DBSCAN) algorithm with a variable semi-supervised ε, which maximizes the distance to the nearest $n-$th earthquake, is proposed. Then, a graph neural network considering the spatial and temporal correlations among clusters is trained for classifying events into three categories: foreshocks, mainshocks and aftershocks. The proposed methodology is applied to the Chilean catalogue of seismic events. Three big clusters representing the earthquakes with a magnitude major than 8.0 on the Ritcher scale are taken for validation. Finally, the results are compared with those obtained by other traditional methods.

C0621: **Bayesian high-dimensional seemingly unrelated regression model with global-local shrinkage**
Presenter: **Dongu Han**, Korea University, Korea, South
Co-authors: Taeryon Choi

Seemingly Unrelated Regression(SUR) is a general framework that can accommodate many useful models, such as multivariate regression or vector autoregressive model. In the era of big data, the number of predictors and the equations to be estimated simultaneously can be both large compared to the sample size. We handle this problem by adopting a variant of horseshoe prior to the parameters. Implementing this prior, we provide an efficient Markov Chain Monte Carlo(MCMC) algorithm without any additional tuning procedures. We also provide some theoretical results, indicating our model works well under mild conditions and provides better estimation compared to the conventional ones. Additionally, we also propose a variational Bayesian method that brings much computational gain without sacrificing the precision of estimation. Several empirical studies show that our proposed method works better than conventional ones.

C0643: **Joint cohort and predictive modelling**
Presenter: **Samuel Emerson**, Durham University, United Kingdom
Co-authors: Louis Aslett

Bayesian logistic regression is a common classification model for binary response data, although in many circumstances the predictive performance and interpretability can be improved with multiple logistic regression models on some partitions of the data. These partitions represent natural clusters in the covariate space where the model may systematically differ. For example, in health modelling settings there may be natural patient cohorts, where interest would often lie in any differences each model reveals between cohorts. We propose a method to jointly find these cohorts and fit the classification model by constructing a graph in covariate space, which is explored by a scheme proposing cuts that form cohorts. A sequential Monte Carlo sampler for the model marginal enables efficient growth and shrinkage of cohorts, making alternatives to logistic regression an easy extension. We discuss associated computational challenges that may arise in large data settings and the work in progress to ameliorate these through principled approximations. There are links to related methods such as a mixture of expert models and model-based clustering.

C0580: **A latent Markov model approach for flexible clustering of longitudinal data**
Presenter: **Zhivko Taushanov**, University of Geneva, Switzerland

Latent Markov models have been successfully used to model and, in some cases, also cluster continuous longitudinal data. We will present a type of approach to longitudinal data, that combines modelling and clustering together. Such a "flexible" clustering would allow some groups to change behaviour over time while keeping them separated from others. The proposed framework consists of a two-level model with a Mixture Transition Distribution in the visible part and a Markovian model driving the latent part, the transition matrix of the latter having a central role. This combined approach would be possible by constraining the transition matrix in a way that it considers data features such as a one- or two-way transition between some pairs of latent states, an impossible transition between others, etc. We will discuss the estimation procedure with its challenges, and possible applications to various domains such as social sciences, demography, psychology etc. Simple examples may include identifying distinct groups while they change temporarily or permanently their behaviour (transition from work to retirement, life-changing events etc.). In such examples, more than one latent state may be associated with the same cluster, aiming to capture changing behaviour of the same group.

CC154 Room Aula D COMPUTATIONAL AND FINANCIAL ECONOMETRICS I **Chair: Thomas Yee**

C0239: **Good contagion: What do networks say about policy transmission**
Presenter: **Kumushoy Abduraimova**, Durham University, United Kingdom

Contagion is frequently perceived as something bad, as a small initial shock amplifying into a systemic crisis, as financial distress propagating from one bank to another, or as a spread of infectious disease. The focus is on good contagion that can facilitate policy propagation. We develop multi-layer network measures of contagion that account for copula and heavy-tailedness structures of the nodes' financial variables. The measures are applied to analyse the European Central Banks interest rate policy transmission. Understanding how network contagiousness, and network structure, more generally, influences the policy transmission is useful for this policy's future successful implementations. The findings indicate that policy transmits most efficiently in severe bearish contagion and least efficiently in intense bullish contagion environments. This result is attributed to the level of attention that markets pay to central bank announcements during turmoil and calm periods. The introduced measures can be used as indicators of systemic importance and as an early warning system of contagion risk.

C0602: **Inflation forecasts disagreement and monetary policy effectiveness**
Presenter: **Dora Xia**, Bank for International Settlements, Switzerland
Co-authors: Sonya Zhu

The term structure of inflation forecasts disagreement is decomposed into disagreement around trend inflation and disagreement around cyclical inflation. While the former has identical impact on forecasts across forecasting horizons, the latter has more muted impact on forecasts at longer-term horizons. We find that the two kinds of disagreement have different impact on monetary policy transmission. Only trend inflation disagreement has a significant impact on monetary policy efficacy. When the trend inflation disagreement is high, monetary policy tightening is ineffective in reducing price pressure.

C0431: **A principal component regression method to incorporate macroeconomic forecasts in modelling expected credit loss**
Presenter: **Helgard Raubenheimer**, North-West University, South Africa
Co-authors: Gerbrand Breed, Tanja Verster

During the financial crisis, the International Accounting Standard Board (IASB) and Financial Accounting Standard Board (FASB) joined their efforts to redesign accounting standards for an improved and simplified expected credit loss (ECL) framework and released the International Financial Reporting Standard (IFRS) 9 in 2014. The quantification of ECL is often broken down into its three components, namely the probability of default (PD), loss given default (LGD) and exposure at default (EAD). The IFRS9 standard requires that the PD model accommodates the influence of the current and the forecasted macroeconomic conditions on default rates. This enables a determination of forward-looking estimates on impairments. A methodology is proposed based on principal component regression (PCR) to adjust IFRS 9 PD term structures for macroeconomic forecasts. We propose that a credit risk index (CRI) is derived from historic defaults to approximate the default behaviour of the portfolio. PCR is used to model the CRI with the macroeconomic variables as the set of explanatory variables. A novice all-subset variable selection is proposed incorporating business decisions. We demonstrate the method's advantages on a real-world banking data set and compare it to several other technics.

C0649: **On the state-space modelling of UK allowance futures prices**
Presenter: **Jun Seok Han**, Macquarie University, Australia
Co-authors: Nino Kordzakhia, Pavel Shevchenko

The term structure of the United Kingdom Allowance (UKA) is modelled using the Schwartz-Smith two-factor model. Futures prices are derived as functions of short-term and long-term factors, leading to a set of simultaneous measurement equations that are used for joint estimation of state variables and model parameters via the Kalman filter and maximum likelihood estimation method. A comparative analysis is performed to compare three reduced-form models, assessing the goodness-of-fit and out-of-sample prediction. We study the price dynamics of historical daily futures prices from May 19, 2021, to March 17, 2022, which matures in December annually.

C0623: **Count data models with endogeneity and selection**
Presenter: **Yves Croissant**, Universite de La Reunion, France

Endogeneity of some covariates in regression and endogenous selection are important problems in econometrics. Relevant methods of estimation have been proposed for linear models, namely instrumental variables, the general method of moments and the Heckman model for sample selection. More recently, these methods have been adapted to non-linear models. The contribution surveys the estimators proposed for count data: Muhally's GMM estimator for endogenous covariates, Terza's ML and NLS estimator for endogenous switching and Greene's ML and NLS estimator for endogenous selection, presents the implementation of these estimators in R and reproduce the original empirical examples contained in the corresponding articles (smoking habits and cigarette demand, credit card holding and major derogatory reports, physician advice and alcohol consumption, car ownership and mobility)

CC216 Room Aula H FUNCTIONAL DATA ANALYSIS **Chair: Dominik Liebl**

C0424: **A Wilcoxon-Mann-Whitney spatial scan statistic for functional data**
Presenter: **Zaineb Smida**, IMAG, university of Montpellier, France

A nonparametric scan method is introduced for functional data indexed in space. The associated scan statistic is derived from the Wilcoxon-Mann-Whitney test statistic defined for infinite dimensional data. It is completely nonparametric as it does not assume any distribution concerning the functional marks. In our simulation study, this scan test appears to be powerful against clustering alternatives. We also apply our method to a data set for extracting features in Spanish province population growth. A significant and relevant spatial cluster with a low rate of demographic change was found in the North-West of Spain.

C0476: **Functional goodness-of-fit tests**

Presenter: **Zdenek Hlavka**, Charles University, Czech Republic

Co-authors: Petr Coupek, Viktor Dolnik, Daniel Hlubinka

The distribution of a Gaussian functional variable (random process) is uniquely determined by the mean function and the covariance operator but it may be characterized also by the so-called characteristic functional (CF). We consider a goodness-of-fit (GOF) test based on a Cramer-von Mises distance between the observed empirical CF and the theoretical CF corresponding to the null hypothesis—in this case, the null hypothesis states the functional observations were generated from a specific family of Gaussian processes. Compared to previously proposed tests of this type, we investigate the functional GOF test also in presence of nuisance parameters, we establish bootstrap consistency, and we discuss the choice of necessary tuning parameters. As an example, we test, e.g., the null hypothesis that the functional observations were generated from an Ornstein-Uhlenbeck process, Vasicek model, or a (fractional) Brownian motion, both with and without unknown parameters, against suitable alternatives. The small sample properties are investigated in a simulation study.

C0637: **Bayesian functional mixed effects model with shape constrained and hierarchical structured gaussian processes**

Presenter: **Jangwon Lee**, Korea University, Korea, South

Co-authors: Taeryon Choi

A Bayesian hierarchical functional mixed-effects model is proposed for grouped data observed unequally spaced time. Our method is formulated as a multivariate functional mixed-effects model whose mean part and random part are modeled by a Bayesian spectral analysis with and without shape constrained, such as monotone, convex, U-shaped, and multiple-extremes. By assuming the hierarchical structure of the spectral coefficient, our model can capture an overall mean trend as well as group trend and subject-specific trend. For flexible modeling for serial dependence in the temporal data, we assume that the error term is a multivariate Ornstein-Uhlenbeck process. The inference is performed by Markov chain Monte Carlo methods.

C0498: **Permutation tests for testing hypotheses in spatial regression model with functional response**

Presenter: **Eva Fiserova**, Palacky University, Czech Republic

Co-authors: Veronika Rimalova, Alessandra Menafoglio, Alessia Pini

The aim is to introduce an approach to hypothesis testing in a functional linear model for spatial data. The proposed method can deal with the spatial structure of data by building a permutation testing procedure on spatially filtered residuals of a spatial regression model. Indeed, due to the spatial dependence existing among the data, the residuals of the regression model are not exchangeable, breaking the basic assumptions of the Freedman and Lane permutation scheme. Instead, it is proposed here to base the permutation test on approximately exchangeable spatially filtered residuals. To evaluate the performance of the proposed method in terms of empirical size and power, a simulation study, examining its behaviour under different covariance settings, is conducted. It will be shown that neglecting the residuals' spatial structure in the permutation scheme (thus permuting the correlated residuals directly) yields a very liberal testing procedure, whereas the proposed procedure based on spatially filtered residuals is close to the nominal size of the test. The methodology will be demonstrated on a real-world data set on the amount of waste production in the Venice province of Italy.

C0520: **Guided structure learning of DAGs for count data**

Presenter: **Thi Kim Hue Nguyen**, University of Padova, Italy

Co-authors: Monica Chiogna, Davide Risso, Erika Banzato

Structure learning of Directed Acyclic Graphs (DAGs) is tackled, with the idea of exploiting available prior knowledge of the domain at hand to guide the search for the best structure. In particular, we assume to know the topological ordering of variables in addition to the given data. We study a new algorithm for learning the structure of DAGs, proving its theoretical consistency in the limit of infinite observations. Furthermore, we experimentally compare the proposed algorithm to a number of popular competitors, in order to study its behavior in finite samples

CC214 Room Aula I ROBUST METHODS II **Chair: Christophe Croux**

C0500: **Robustness and outlier detection of Bayesian model residuals with mixtures of normal, heavy-tailed and skewed components**

Presenter: **Alexandra Posekany**, University of Technology Vienna, Austria

Outliers and skewed or heavy-tailed data frequently occur in data analytical problems in many fields. We consider notions of Bayesian robustness for various model types and compare them against classical robust estimators. Three aspects form the basis of Bayesian robustness: prior, likelihood and loss robustness. Generally, a considerate Bayesian analysis considers prior robustness through a sensitivity analysis, varying hyper-parameters and checking their influence. Loss functions connect with classical notions of robustness, e.g. reporting the posterior's median rather than mean. Yet, they are often disregarded, as estimating means is the basis of Monte Carlo simulation. The final notion of Bayesian robustness is robustifying the likelihood. Constructing normally distributed likelihood models is often due to computational convenience. We wish to provide a robust estimation of parameters of the main part of the data through a normal or skewed distribution as likelihood, while simultaneously identifying the outlying part of the data represented by one or more skewed or heavy-tailed mixture components. Through the component labels and posterior weights, we can identify the noisy or outlying parts of the data for filtering or inspecting the data quality.

C0538: **Robust Pitman type estimators for moment condition models**

Presenter: **Aida Toma**, Bucharest University of Economic Studies, Romania

Co-authors: Amor Keziou, Luiza Badin, Silvia Dedu

Robust minimum empirical divergence estimators are presented for moment condition models, based on truncated orthogonality functions and dual forms of divergences. For moment condition models invariant with respect to additive or multiplicative transformations groups, these estimators are also equivariant. For models invariant with respect to additive groups, robust Pitman-type estimators are proposed. Some examples based on Monte Carlo simulations illustrate the performance of the estimation method.

C0640: **Accounting for asymmetry in M-estimation: A Julia package**

Presenter: **Manuel Stapper**, WWU Muenster, Germany

A software package is presented that enables the user to carry out M-Estimation in different univariate settings. Besides location estimation, it provides methods for parameter estimation for i.i.d. samples, regression, and fitting time series models. A focus is put on asymmetric distributions, in which estimates are potentially biased when using symmetric loss functions. It is accounted for in two ways: an established consistency correction and an adaptive estimation procedure based on asymmetric loss functions. The latter enables the user to estimate parameters of an i.i.d. sample with a selected relative asymptotic efficiency compared to Maximum Likelihood Estimation. The package includes not only four frequently used loss functions (Huber, Tukey's Biweight, Hampel and Andrew's Wave) but also allows applying the methods with self-defined loss functions. By utilising Julia's Multiple Dispatch, parameter estimates are computed fast for more than 30 common distributions in the Distributions.jl package. Fallback functions allow carrying out parameter estimation for any other univariate distribution with a known probability (density) function. The extension to time series applications is discussed and showcased by fitting a count data model to real-world data.

C0258: **The penalized robust double exponential estimators**
Presenter: **Jolien Ponnet**, KU Leuven, Belgium
Co-authors: Pieter Segaert, Stefan Van Aelst, Tim Verdonck

The family of double exponential distributions models both the mean and the dispersion as a function of covariates in the generalized linear model (GLM) framework. Since standard maximum likelihood inference is highly susceptible to the possible presence of outliers, we propose the robust double exponential (RDE) estimator. We focus on the penalized versions of the RDE estimator. First of all, we consider penalties for obtaining sparsity in high-dimensional settings. This allows us to select the most important predictors out of a large number that may even exceed the sample size. Secondly, we consider regularization penalties in the context of flexible smooth estimation via generalized additive models (GAMs). Hereby, the GLM for the mean and/or dispersion is replaced by a GAM. Simulation studies demonstrate the decent and robust performance of both penalized RDE estimators. Finally, the penalized RDE estimators are illustrated on real data sets.

CC221 Room Aula Q DIMENSION REDUCTION **Chair: Matteo Farne**

C0399: **Supervised component-based generalized linear regression with conditionally covarying responses**
Presenter: **Julien Gibaud**, IMAG, France
Co-authors: Xavier Bry, Catherine Trottier

Originally, the Supervised Component-based Generalized Linear Regression (SCGLR) was designed to find explanatory components in a large set of possibly highly redundant covariates. This methodology optimizes a trade-off criterion between the model's Goodness-of-Fit and some Structural Relevance of directions with respect to the explanatory variables. This methodology allows both to find strong explanatory directions and to produce regularized predictors in a high-dimensional framework. Later on, SCGLR was extended with the aim to search for components in a thematic partitioning of the explanatory variables. However, SCGLR assumed the responses to be independent conditional on the explanatory covariates, which is not often realistic. To overcome this limitation, we propose to extend SCGLR by modeling the responses' conditional variance-covariance matrix using a small number of latent random variables called factors. More formally, a response matrix is assumed to depend, through a Generalized Linear Model, on a set of explanatory variables partitioned into several conceptually homogenous variable groups, viewed as explanatory themes and an unknown number of common latent factors accounting for their dependence structure. The method is tested on simulated data and then applied to a floristic ecology dataset.

C0551: **An empirical study on nonlinear structure extraction with measures of dependence**
Presenter: **Shoma Ishimoto**, Hokkaido University, Japan
Co-authors: Hiroyuki Minami, Masahiro Mizuta

A method is proposed for extracting nonlinear structure from multi-dimensional data by extending dimension reduction with measures of dependence; various measures of dependence have been proposed to evaluate the strength of linear or nonlinear relationships between 2 variables. To achieve our goal, we adopt the measures in place of the indices in popular dimension reduction, and find the directions that maximize them. We apply our idea to numerical examples with typical nonlinear structures (quadratic, cubic, sinusoid, circle, power) with random noise. We introduce 7 types of measures of dependence (MIC, TIC, Hoeffding's D, distance correlation, KSG Estimator, HSIC, RDC) and project the simulated data into lower space according to the measures. Through 100 times simulation, we discuss the results from the viewpoints of performance, specific features and our competence. We conclude the most suitable one to meet our idea.

C0558: **ALS algorithm for CDPCA on high-dimensional data sets: An empirical study**
Presenter: **Adelaide Freitas**, University of Aveiro, Portugal
Co-authors: Maurizio Vichi

Applied on high-dimensional data sets, constrained Principal Component Analysis (PCA) techniques, those yielding sparse solutions, are particularly useful to make easier the interpretation of the components. Clustering and Disjoint Principal Component Analysis (CDPCA) is a constrained PCA that promotes sparsity in the loadings matrix and, simultaneously, the identification of clusters of objects. Based on simulated and real gene expression data sets where the number of variables is higher than the number of the objects, we empirically evaluate the performance of the Alternating Least Square (ALS) algorithm, a heuristic iterative procedure proposed in the specialized literature to perform CDPCA. Our numerical tests show that ALS performs well and produces satisfactory results in terms of solution precision. In recovering the true object clusters, the complexity of the data structure (i.e., the error level of the CDPCA model on which the data was generated) seems to influence the ability of ALS when the sample size is not so high. For a lower sample size, ALS performs better when the error level is lower. The proportion of explained variance by the components estimated by ALS is affected by the data structure complexity (the higher the error level, the lower variance).

C0548: **Anomaly detection with kernel density estimation on manifolds**
Presenter: **Fan Cheng**, Monash University, Australia
Co-authors: Anastasios Panagiotelis, Rob Hyndman

Manifold learning can be used to obtain a low-dimensional representation of the underlying manifold given the high-dimensional data. However, kernel density estimates of the low-dimensional embedding with a fixed bandwidth fail to account for the way manifold learning algorithms distort the geometry of the underlying Riemannian manifold. We propose a novel kernel density estimator for any manifold learning embedding by introducing the estimated Riemannian metric of the manifold as the variable bandwidth matrix for each point. The geometric information of the manifold guarantees a more accurate density estimation of the true manifold, which subsequently could be used for anomaly detection. To compare our proposed estimator with a fixed-bandwidth kernel density estimator, we run two simulations with 2-D metadata mapped into a 3-D swiss roll or twin peaks shape and a 5-D semi-hypersphere mapped in a 100-D space, and demonstrate that the proposed estimator could improve the density estimates given a good manifold learning embedding and has higher rank correlations between the true and estimated manifold density. A shiny app in R is also developed for various simulation scenarios. The proposed method is applied to density estimation in statistical manifolds of electricity usage with the Irish smart meter data. This demonstrates our estimator's capability to fix the distortion of the manifold geometry and to be further used for anomaly detection in high-dimensional data.

C0403: **Probabilistic principal curves on Riemannian manifolds**
Presenter: **Seungwoo Kang**, Seoul National University, Korea, South
Co-authors: Hee-Seok Oh

A new curve fitting approach is studied that is useful for the representation and dimension reduction of data on Riemannian manifolds. We extend the probabilistic formulation of the curve passing through the middle of data on Euclidean space to Riemannian symmetric space. To this end, we define a principal curve based on a mixture model for observations and unobserved latent variables, and propose a new algorithm to estimate the principal curve for given data points on Riemannian manifolds using a series of procedures in 'unrolling, unwrapping, and wrapping' and EM algorithm. Some properties for justification of the estimation algorithm are further investigated. Results from numerical examples, including several simulation sets on hyperbolic space, sphere, special orthogonal group, and a real data example, demonstrate the promising empirical properties of the proposed probabilistic approach.

CC203 Room Aula F STATISTICS AND DATA SCIENCE **Chair: Frank Miller**

C0581: **Distribution of a linear combination of generalized logistic random variables with application to financial returns**

Presenter: **Andjela Mijanovic**, University of Montenegro, Montenegro

The cumulative distribution function for n independent generalized logistic random variables in terms of the Fox H-function is derived. The exact expressions for evaluating the probability density function, and cumulative distribution function of a linear combination of n independent generalized logistic random variables are derived by using Mellin and inverse Mellin transform, also a method based on the numerical inversion of the characteristic function is considered. We compare the numerical precision and efficiency of the considered methods. Application of the considered linear combination in the field of financial returns will be presented.

C0593: **Familial inference**

Presenter: **Ryan Thompson**, Monash University, Australia

Co-authors: Catherine Forbes, Steven MacEachern, Mario Peruggia

Statistical hypotheses are translations of scientific hypotheses into statements about one or more distributions, often concerning their center. Tests that assess statistical hypotheses of center implicitly assume a specific center, e.g., the mean or median. Yet, scientific hypotheses do not always specify a particular center. This ambiguity leaves the possibility of a gap between scientific theory and statistical practice that can lead to rejection of a true null. In the face of replicability crises in many scientific disciplines, "significant results" of this kind are concerning. Rather than testing a single center, we propose testing a family of plausible center", such as that induced by the Huber loss function (the "Huber family"). Each center in the family generates a testing problem, and the resulting family of hypotheses constitutes a familial hypothesis. A Bayesian nonparametric procedure is devised to test familial hypotheses, enabled by a novel pathwise optimization routine to fit the Huber family. We verify the favorable properties of the new test through numerical simulation in one- and two-sample settings. Two experiments from psychology serve as real-world case studies.

C0510: **On some issues related to the fairness of algorithms**

Presenter: **Gilbert Saporta**, CNAM, France

Fairness of algorithms is the subject of a large body of literature, guides, computer codes and tools. Machine Learning and AI algorithms commonly used to accept loan applications, select responses to job offers, etc. are often accused of discriminating against groups. We will begin by examining the relationship between fairness, explainability, and interpretability. One might think that it is better to understand how an algorithm works in order to know whether it is fair, but in fact, this is not the case, because transparency or explainability are relative to the algorithm, whereas fairness concerns its differential application to groups of individuals. There is a wide variety of often incompatible measures of fairness. Moreover, questions of robustness and precision are often ignored. The choice of a measure is not only a matter of statistical considerations but of ethical choices. The biases so-called of the algorithms are often only the reproduction of those of previous decisions found in the training data. But they are not the only ones. We will attempt to draw up a typology of the main biases: statistical, societal, cognitive, etc. and discuss the links with causal models.

C0271: **A statistical learning view of simple kriging**

Presenter: **Emilia Siviero**, Telecom Paris, France

Co-authors: Emilie Chautru, Stephan Clemencon

In the Big Data era, massive datasets exhibiting a possibly complex spatial dependence structure are becoming increasingly available. The standard probabilistic theory of statistical learning does not apply directly, and guarantees of the generalization capacity of predictive rules learned from such data are left to establish. We analyze here the simple Kriging task, the flagship problem in Geostatistics: the values of a square-integrable random field $X = \{X_s\}_{s \in S}$, $S \subset \mathbb{R}^2$, with unknown covariance structure are to be predicted with minimum quadratic risk, based upon observing a single realization of the spatial process at a finite number of locations $s_1, \ldots, s_n$ in S. Despite the connection of this minimization problem with kernel ridge regression, establishing the generalization capacity of empirical risk minimizers is far from straightforward, due to the non-i.i.d. nature of the spatial data $X_{s_1}, \ldots, X_{s_n}$ involved. Nonasymptotic bounds of order $O_{\mathbb{P}}(1/n)$ are proved for the excess risk of a plug-in predictive rule mimicking the true minimizer in the case of isotropic stationary Gaussian processes observed at locations forming a regular grid. These theoretical results, as well as the role played by the technical conditions required to establish them, are illustrated by various numerical experiments and hopefully pave the way for further developments in statistical learning based on spatial data.

C0677: **Fractals in time series data: The methodological case for fractional differencing and power spectral density approaches**

Presenter: **Matthijs Koopmans**, Mercy College, USA, United States

The question of whether time series data contain fractal patterns is of interest because of their non-randomness. Therefore, they need to be accounted for in the modeling stage of the analysis. Moreover, fractals are of substantive interest, as they indicate self-similarity and scale invariance. Many statistical techniques are available to detect whether time series data contain fractal patterns, including detrended fluctuation analysis, re-scaled range analysis, fractional differencing (time domain) and a variety of spectral regression approaches (frequency domain that fit linear functions to log-log power spectra). There is more variability than there should be in the fractality estimators produced by these techniques, as well as great variation in their ability to distinguish fractal variance from seasonal and short-range dependencies. This presentation uses three real datasets (river Nile flow, daily school attendance, daily birth to teens recordings), and a set of simulations with varying levels and types of short and log-range dependency to show that fractional differencing and power spectral density analysis is superior to the other techniques, provided that they are used in conjunction. The former because of its use of stepwise model comparisons to distinguish fractal from non-fractal patterns, and the latter because its power spectrums visually demonstrate scale invariance.

Thursday 25.08.2022 14:15 - 15:45 Parallel Session K – COMPSTAT2022

CV226 Room Aula G CLUSTERING AND CLASSIFICATION II (VIRTUAL) **Chair: Marco Bee**

C0322: ***k*-means cluster analysis: A study on cervical cancer mortality in Veracruz, Mexico**
Presenter: **Monserrat Martinez de los Santos**, Universidad Veracruzana, Mexico
Co-authors: Emmanuel Morales-Garcia, Candy Obdulia Sosa Jimenez, Maribel Carmona Garcia

Cervical Cancer is a public health problem worldwide, given that there are large numbers of deaths due to this disease. In the Mexican Republic, the rates are high. The state of Veracruz is no exception and this condition has occupied one of the first mortality reasons. $k-$means cluster has been used to classify the regions of Veracruz in order to learn about the mortality produced by this disease. Data from the cnegsr have been used. The first cluster obtained is characterized by women with complete basic education, who had basic medical service, aged between 40 and 60 years and belonging to the region, Olmec and capital (1). While cluster two presents women between 50 and 89 years old, 59% had some medical service, with incomplete basic education. They belong to the regions, Olmec, Capital Mountains and Totonacapan. Finally, cluster three contained women between 45 and 79 years, basic education to higher, with medical insurance, mainly from mountains, Olmec, leeward and the capital.

C0642: **LEFDA: An extension of the classical LDA**
Presenter: **Alice Giampino**, University of Milano-Bicocca, Italy
Co-authors: Roberto Ascari, Sonia Migliorati

Latent Dirichlet Allocation (LDA) is a popular statistical tool for the analysis of text documents when the goal is detecting latent topics. A well-known limitation of the LDA is its inability to model positive correlations between topics. This is attributable to the stiffness of the Dirichlet distribution, which is the standard prior for the topic distributions. The aim is to perform a preliminary study of the extended flexible Dirichlet (EFD) as an alternative prior. The latter is a generalization of the Dirichlet distribution defined as a particular structured mixture allowing for positive correlations between its elements. The EFD distribution retains many good theoretical properties of the Dirichlet one, such as identifiability and also explicit expressions of joint moments and closure under many relevant operations on the simplex. Furthermore, the introduction of additional parameters establishes more flexibility, while still maintaining the interpretability of the model, as well as conjugacy with respect to the multinomial model. The generalization of the LDA based on the EFD distribution is illustrated via an application to real data using Markov Chain Monte Carlo (MCMC) methods.

C0668: **Using neural clustering in spatial and non spatial models**
Presenter: **Jean-Charles Lamirel**, LORIA, France
Co-authors: Cecile Hardouin

The aim of spatial econometrics is to analyze and/or predict the relationship between one dependent variable Y with other variables, taking into account spatial dependence. In the framework of Spatial Autoregressive Models (SAR), Y is linked to covariates and to the values of its own adjacent values via a neighbourhood matrix $W = (w_{ij})$. Weights w_{ij} are usually linked to the geographical distances between the locations where the data were collected from. Considering that similar values of the dependent variable can result from geographical proximity, but also from the similarity of the variables, we propose weights based on neural clustering. Kohonen SOM or Growing Neural Gas algorithms provide distances between nodes that we use directly to define weights w_{ij}. In the general spatial model setting, we need two neighbourhood matrices and the difficulty rising then is to find a second matrix; this issue is simply solved by using neural distances. The results we obtained are at least as well as the ones obtained from the geographical distances-based design. This approach can be generalized to non-geographical data; SAR models with neural distances design can be used to model any data set, non necessarily geographically referenced. We illustrate our approach with the study of real data sets.

C0446: **Parsimonious seemingly unrelated linear cluster-weighted models for contaminated data**
Presenter: **Gabriele Perrone**, Department of Statistical Sciences, University of Bologna, Italy
Co-authors: Gabriele Soffritti

Nowadays, in several areas, it is frequent for a researcher to be involved in the task of systematically extracting information from data sets that are too large or complex to be dealt with by traditional statistical methods. This is also true in multivariate linear regression analysis when samples are characterised by unobserved heterogeneity. A modern approach to this problem is represented by the Gaussian linear cluster-weighted models, which allow to simultaneously perform model-based cluster analysis and multivariate linear regression analysis with random predictors. Robustified models have been recently developed, based on the use of the contaminated Gaussian distribution, which can manage the presence of mildly atypical observations. A more flexible class of contaminated Gaussian linear cluster-weighted models is specified, in which the researcher is free to use a different vector of covariates for each response. The novel class also includes parsimonious models, where parsimony is attained by imposing suitable constraints on the component-covariance matrices of either the responses or the covariates. Identifiability conditions are illustrated and discussed. An expectation-conditional maximisation algorithm is provided for the maximum likelihood estimation of the model parameter. The effectiveness and usefulness of the proposed models are shown through the analysis of simulated and real datasets.

CI005 Room Aula F ROBUST STATISTICS **Chair: Peter Filzmoser**

C0275: **The influence function of graphical lasso estimators**
Presenter: **Ines Wilms**, Maastricht University, Netherlands
Co-authors: Gaetan Louvet, Jakob Raymaekers, Germain Van Bever

Graphical models are nowadays often estimated using regularization that is aimed at reducing the number of edges in a network. By relying on edge-sparsity as a simplifying structure, the conditional dependency network among (potentially a large number of) variables can then be presented in a compact manner. The Graphical Lasso (Glasso) is a common choice to obtain such sparse graphical models. Glasso lacks, however, robustness to outliers. To overcome this problem, one typically applies a robust plug-in procedure where the Glasso is computed from, for instance, an initial pairwise robust covariance/correlation estimate instead of the classical sample covariance estimate, thereby providing protection against outliers. We derive and compare the influence function of the classical Glasso to various robustified versions, as well as their corresponding asymptotic variances. Simulation results provide further insights into their finite sample performance.

C0361: **Scalable robust estimators for non-parametric regression models**
Presenter: **Matias Salibian-Barrera**, The University of British Columbia, Canada
Co-authors: Xiaomeng Ju

Many robust estimators for nonparametric regression models have been proposed in the literature. Unfortunately, most non-parametric regression estimators generally do not scale well when the number of explanatory variables is relatively large (c.f. the curse of dimensionality). Additive models can avoid this problem, at the expense of having to impose a strong structure in the regression function. A different family of non-parametric regression estimators is given by gradient boosting, which constructs a regression predictor using a linear combination of simple base learners (e.g. regression trees), which can be used effectively even with many covariates. A robust variant of gradient boosting for regression problems can be obtained by minimizing a properly defined loss function. To avoid relying on ad-hoc estimates of the residual scale that change in every iteration, we use a two-stage approach (as with MM-estimators for linear regression): we first minimize a robust residual scale estimator, and then improve it by optimizing an M-type loss function. Simulation studies and several data analyses show that, when no atypical observations are present, the robust boosting approach works as well as the standard gradient boosting one with a squared loss. As expected, when the data contain outliers the robust boosting estimator outperforms existing alternatives.

C0608: **Sparse regression for large data sets with outliers**
Presenter: **Christophe Croux**, Edhec Business School, France
Co-authors: Ines Wilms, Lea Bottmer

The linear regression model remains an important workhorse for data scientists. However, many data sets contain many more predictors than observations. Besides, outliers, or anomalies, frequently occur. An algorithm is proposed for regression analysis that addresses these features typical for big data sets. The resulting regression coefficients are sparse, meaning that many of them are set to zero, hereby selecting the most relevant predictors. A distinct feature of the method is its robustness with respect to outliers in the cells of the data matrix. The excellent performance of this robust variable selection and prediction method is shown in a simulation study. A real data application on car fuel consumption demonstrates its usefulness.

CO047 Room Aula B STATISTICAL METHODS FOR STATISTICALLY CHALLENGING DATA (VIRTUAL)	**Chair: Anuradha Roy**

C0156: **From the analysis of the composite indicators to the analysis of the symbolic composite indicators**
Presenter: **Carlo Drago**, University of Rome Niccolo Cusano, Italy

Nowadays, composite indicators represent and summarize complex phenomena that cannot be measured by looking at a single variable. In this sense, the advantage of using composite indicators is that they can be used directly in policy analysis. However, composite indicators are based on assumptions that may vary and are subjective (the treatment of missing data, weighting, the choice of variables to use, etc.). In this context, we propose using symbolic composite indicators, which are more informative than the original composite indicators since they explicitly internalize or take into account the variability of the original composite indicators when considering the different relevant choices on the factors. In this sense, sensitivity analysis is directly incorporated into the construction of the indicator. It could be considered possible to interpret the parameters of the final indicators.

C0279: **Analytical tools for whole-brain networks: Fusing statistics and network science to understand brain function**
Presenter: **Sean Simpson**, Wake Forest University School of Medicine, United States
Co-authors: Mohsen Bahrami, Chal Tomlinson, Paul Laurienti

Brain network analyses have exploded in recent years, and hold great potential in helping us understand normal and abnormal brain function. Network science approaches have facilitated these analyses and our understanding of how the brain is structurally and functionally organized. However, the development of statistical methods that allow relating this organization to health outcomes has lagged behind. We have attempted to address this need by developing analytical tools that allow relating system-level properties of brain networks to outcomes of interest. These tools serve as synergistic fusions of statistical approaches with network science methods, providing needed analytic foundations for whole-brain network data. We delineate two recent approaches–a mixed-modeling framework for dynamic network analysis and a regression framework for relating distances between brain network features to covariates of interest–that expand the suite of analytical tools for whole-brain networks and aid in providing complementary insight into brain function.

C0293: **Bayesian analysis of multivariate linear mixed models with censored and missing responses**
Presenter: **Wan-Lun Wang**, National Cheng Kung University, Taiwan

Multivariate longitudinal data usually exhibit complex features such as the presence of censored responses due to detection limits of the assay and unavoidable missing values arising when participants make irregular visits that lead to intermittently recorded characteristics. A generalization of the multivariate linear mixed model constructed by taking impacts of censored and intermittent missing responses into account simultaneously, which is named the MLMM-CM, has been recently proposed for more precisely analyzing such kinds of data. The aim is at presenting a fully Bayesian approach to the MLMM-CM for addressing the uncertainties of censored and missing responses as well as unknown parameters. Bayesian computational techniques based on the inverse Bayes formulas (IBF) coupled with the Gibbs scheme are developed for carrying out posterior inference of the model. The proposed methodology is illustrated through a simulation study and a real-data example from the Adult AIDS Clinical Trials Group 388 study. Numerical results show empirically that the proposed Bayesian methodology performs satisfactorily and offers reliable posterior inference.

C0324: **Model-based clustering via mixtures of unrestricted skew normal factor analyzers with missing values**
Presenter: **Tsung-I Lin**, National Chung Hsing University, Taiwan

Mixtures of factor analyzers (MFA) based on the restricted skew normal distribution (rMSN) have been shown to be a flexible tool to handle asymmetrical high-dimensional data with heterogeneity. However, the rMSN distribution is oft-criticized a lack of sufficient ability to accommodate potential skewness arising from more than one feature space. An alternative extension of MFA is presented by assuming the unrestricted skew normal (uMSN) distribution for the component factors. In particular, the proposed mixtures of unrestricted skew normal factor analyzers (MuSNFA) can simultaneously capture multiple directions of skewness and deal with the occurrence of missing values or nonresponses. Under the missing at random (MAR) mechanism, we develop a computationally feasible expectation conditional maximization (ECM) algorithm for computing the maximum likelihood estimates of model parameters. Practical aspects related to model-based clustering, prediction of factor scores and missing values are also discussed. The utility of the proposed methodology is illustrated with the analysis of simulated data and the Pima Indian women's diabetes data containing genuine missing values.

CO043 Room Aula D ASSOCIATION, DEPENDENCE AND COPULAS	**Chair: Sebastian Fuchs**

C0183: **Analysing the relationship between district heating demand and weather conditions through conditional mixture copula**
Presenter: **F Marta L Di Lascio**, Free University of Bozen-Bolzano, Italy
Co-authors: Andrea Menapace, Maurizio Righetti

Efficient energy production and distribution systems are urgently needed to reduce world climate change. Modern district heating systems play a crucial role in this context since they are energy distribution services that exploit renewable sources and use smart grids for any heat request in the urban area. To enhance the heat production schedule, in-depth knowledge of thermal energy demand, which is mainly affected by weather conditions, is essential. We hence propose a mixture copula-based approach to investigate the complex relationship between meteorological variables, such as outdoor temperature and solar radiation, and thermal energy demand in the district heating system of the Italian city Bozen-Bolzano. We analyse data collected from 2014 to 2017 and estimate copulas after removing serial dependence in each time series using autoregressive integrated moving average models. A finite mixture of heterogeneous parametric copulas to generate dependence structures not expressible by the existing models is specified. Precisely, we selected a mixture of an unstructured Student-t copula and a flipped Clayton copula that makes it possible to differentiate the magnitude of dependence in each tail and to exhibit both heavy tail and asymmetric dependence. We derive the conditional copula-based probability function of thermal energy demand given meteorological variables and provide useful insight on efficiently planning the heat production and distribution.

C0488: **Total positivity of copulas from a Markov kernel perspective**
Presenter: **Marco Tschimpke**, Paris Lodron University Salzburg, Austria
Co-authors: Sebastian Fuchs

An alternative way to examine dependence structures consists in checking whether a copula fulfils certain (positive) dependence properties such as stochastically increasingness (SI) or total positivity of order 2. We investigate total positivity of order 2 for a copula's Markov kernel (MK-TP2 for short), presumably the strongest dependence property defined for any copula. We mainly examine the MK-TP2 property within the class of Archimedean and Extreme-Value copulas and characterise the dependence property based on the respective generators.

C0521: **Statistical copulas approach for dependence in remote sensing problems**
Presenter: **Cristiano Tamborrino**, University of Bari, Italy

Remote sensing data acquired from Earth by satellites and airplanes have become increasingly important in various environmental and ecological contexts (e.g. agriculture, oceanographic or urban areas) over the last decade. The constellations of satellites and sensors on board are numerous and with ever-increasing technologies. This allows the acquisition of Hyperspectral (HS), Multispectral (MS) and Synthetic Aperture Radar (SAR) images with a very high spatial and spectral resolution, moreover, it is possible to collect a large amount of historical data, with a very short time interval, in different areas of the planet. The analysis of this large amount of data requires ever more precise and fast methods that must take into account not only the dependence on the spectral characteristics of every single image, but also on the temporal ones. Copulas are an excellent statistical tool, capable of modeling joint distributions between any random variable. Recently, it has been seen that the use of copulas alongside the classic machine learning algorithms for classification, clustering or anomaly detection leads to a more precise and robust quantitative analysis even with respect to the latest developments with Neural Network architectures. We will apply this tool in different areas of remote sensing data analysis, the proposed approaches will be tested on hyperspectral and multispectral images and the results are compared with the most advanced methods.

C0441: **Using feature selection based on multivariate statistical dependence for churn prediction in the automotive industry**
Presenter: **Thimo Kasper**, University of Salzburg, Austria
Co-authors: Laura Koenig, Markus Gruber, Thomas Soboll, Wolfgang Trutschnig

In recent years customer retention and preventing customer churn is of increasing importance for businesses in various industries. Particularly in the non-contractual automotive aftersales market, where churn is a not directly observable latent event, retaining customers prone to defecting is key for the profitability of dealerships and workshops. Therefore, working with real-life data from a group of workshops/dealerships from an international automotive distributor in an Austrian region, we tackle the question of how to assess aftersale customer churn probabilities by using techniques from statistics and machine learning in order to allow for dynamic customer selection and targeted retention marketing. Driven by the need for efficient, well interpretable models, special focus is assigned to the feature selection procedure - four recently developed methods based on bivariate and multivariate statistical dependence are benchmarked against random forest's feature importance and a selection based upon Pearson's correlation coefficient. Our findings show that reliable, well-performing customer churn prediction with low model complexity is indeed possible in the context of automotive aftersales.

CO053 Room Aula H NON-PROPORTIONAL HAZARDS IN SURVIVAL DATA Chair: Francesca Gasperoni

C0272: **Survival analysis under non-proportional hazards: Investigating non-inferiority or equivalence in time-to-event data**
Presenter: **Kathrin Moellenhoff**, Heinrich Heine University Dusseldorf, Germany
Co-authors: Achim Tresch

Time-to-event outcomes are frequently observed in medical research, for instance, in the area of oncology or cardiovascular diseases. A commonly addressed issue is the comparison of a test to a reference treatment regarding survival. For this purpose, an analysis based on Kaplan-Meier curves, followed by a log-rank test, is still the most popular approach. In case of addressing non-inferiority or equivalence, extensions of the log-rank test are used. Using one of these approaches, a direct interpretation is obtained by summarizing the treatment effect in one single parameter, given by the hazard ratio of the two treatments, assumed to be constant over time. However, in numerous trials, hazards are non-proportional, and these approaches suffer from a loss of power. We propose a parametric framework to assess equivalence or non-inferiority for survival data. Assuming various time-to-event distributions, we first derive pointwise confidence bands for both, the hazard ratio and the difference in the survival curves. Second, we perform a test addressing non-inferiority and equivalence by directly comparing the survival functions at certain time points or over an entire time interval. We demonstrate the validity of the approach even in settings where sample sizes are small.

C0360: **The differences of restricted mean survival time curves estimated using pseudo-values**
Presenter: **Federico Ambrogi**, University of Milan, Italy
Co-authors: Simona Iacobelli, Per Kragh Andersen

Hazard ratios are ubiquitously used in time-to-event analysis to quantify treatment effects. Although hazard ratios are invaluable for hypothesis testing, other measures of association, both relative and absolute, may be used to fully elucidate study results. Restricted mean survival time differences between groups have been advocated as useful measures of association. Recent work focused on model-free estimates of the difference in restricted mean survival for all follow-up times instead of focusing on a single time horizon. The resulting curve can be used to quantify the association in time units with a simultaneous confidence band. A model-based alternative is proposed with estimation using pseudo-values easily implementable with available software. It is also possible to compute a confidence region for the curve. As a by-product, the "time until treatment equipoise" (TUTE) is also studied. Examples with crossing survival curves will be used to illustrate the different methods together with some simulations.

C0413: **GFDsurv: A flexible toolbox to analyse nonproportional hazards in factorial survival designs**
Presenter: **Marc Ditzhaus**, Otto-von-Guericke University Magdeburg, Germany
Co-authors: Dennis Dobler, Arnold Janssen, Markus Pauly

While the log-rank test and hazard ratios were the gold standard in time-to-event analysis for a long time, there is a recent trend towards alternative methods not relying on the proportional hazard assumption. The reason for this change is violations of the proportional hazard assumption frequently observed in real data, among others, in oncology. To tackle this, we developed different strategies to handle non-proportional hazards in factorial designs and implemented them in the R-package GFDsurv including a user-friendly shiny app. These strategies cover a nonparametric approach and procedures based on novel estimands, such as concordance probabilities, survival medians and, in the near future, restricted mean survival times. We present the methodology behind this flexible toolbox including appropriate resampling strategies for a better small sample size performance. Moreover, its application is illustrated by analysing a recent study on asthma, for which the assumption of proportional hazards is not justifiable.

C0586: **Stratified weighted log-rank tests in settings with anticipated delayed effects**
Presenter: **Jose Jimenez**, Novartis, Switzerland
Co-authors: Dominic Magirr

Delayed separation of survival curves is a common occurrence in confirmatory studies in immuno-oncology. Many novel statistical methods that aim to efficiently capture potential long-term survival improvements have been proposed in recent years. However, the vast majority do not consider stratification, which is a major limitation considering that most (if not all) large confirmatory studies currently employ a stratified primary analysis. We combine recently proposed weighted log-rank tests that have been designed to work well under a delayed separation of survival curves, with stratification by a baseline variable. The aim is to increase the efficiency of the test when the stratifying variable is highly prognostic for survival. As there are many potential ways to combine the two techniques, we compare several possibilities in an extensive simulation study. We also apply the techniques retrospectively to two recent randomized clinical trials.

CO178 Room Aula Q COMPUTATIONAL STATISTICS FROM THE LENS OF YOUNG RESEARCHERS I Chair: Marta Disegna

C0493: **Dependence analysis of aggregate zonal imbalance in the Italian electricity market**
Presenter: **Aurora Gatto**, University of Salento, Italy
Co-authors: Fabrizio Durante, Francesco Ravazzolo

The purpose is to analyse the features and the dynamics of the Italian aggregate zonal electricity imbalance, that is the algebraic sum, changed in sign, of the amount of electricity procured by the Italian national operator from a given Italian electricity macro-zone. In particular, the problem of determining possible correlations and dependencies with the aggregate zonal imbalance in electricity markets is considered. A copula-based model is provided in order to understand the dependence and association between the aggregate zonal imbalance and other variables of interest such as forecasted demand, forecasted wind and solar PV generation. Thanks to the flexibility of this approach, we have identified the non-linear linkage between the aggregate zonal imbalance and the other variables of interest by means of a multivariate approach.

C0341: **Machine learning-based sentiment analysis with fuzzy data to predict online customer satisfaction**
Presenter: **Nicolo Biasetton**, Universita degli Studi di Padova, Italy
Co-authors: Rosa Arboretti, Elena Barzizza, Riccardo Ceccato, Marta Disegna, Luca Pegoraro, Luigi Salmaso

Big Data and Web 2.0 allow gathering a huge amount of free and timely online reviews that customers write on a variety of products/services. Generally, review web platforms ask customers to leave a textual review along with rates regarding the overall product/service and its key aspects. Most of the studies adopting ML-based SA use the general rating as an independent variable, and some of them also include aspect rating within the application. This type of approach makes it possible to predict the general rate through the specific rates collected on product/service aspects, if available, and textual reviews. However, despite being a user-friendly, easy-to-develop and to-administer instrument, Likert-type scales used to collect rating data are unprecise tools which generate ordinal variables that cannot be analysed by statistical methods defined on a metric space. In fact, the distance between two consecutive items cannot be either defined or presumed equal. In such a context, fuzzy theory can be used to recode customers' rates into fuzzy numbers before the adoption of a suitable ML algorithm for fuzzy data. This procedure allows the obtention of more precise predictions of general customer satisfaction. Our approach is presented and discussed using real data, highlighting its main advantages.

C0349: **A novel application of spatial statistics in clustering the world's diets**
Presenter: **Thai Le**, University of Economics and Business, Vietnam National University Hanoi, Vietnam

Notwithstanding the nascent literature on spatial clustering in food economics, previous studies have largely ignored the spatial dimension in clustering dietary patterns. The application of a novel Copula-based K-Medoids Fuzzy Space-Time (COFUST) clustering algorithm is presented for identifying agglomerations of countries that are characterised by similar past trends of food consumption, taking into account their spatial relationship. Specifically, we employ the calorie availability series for 118 countries over the period 1961 to 2013. A key advantage of this approach is the ability to examine the role of the space as a contextual factor for dietary behaviour. The identified clusters not only show similarity in the calorie trajectories but also share homogeneous environment conditions of food consumption. A great novelty is the utilisation of an economic proximity measure instead of traditional metrics for the geographical space. Finally, we introduce the Generalised Fuzzy Morans index that measures the spatio-temporal autocorrelation for spatial units that are collected over time. This index could assist in the selection of the optimal spatial coefficient in the clustering procedure. Results using both simulated and real data show that ignoring the spatial relationship can lead to incorrect interpretation of the clustering results.

C0423: **On modelling and estimating geo-referenced count spatial data with excessive zeros**
Presenter: **Diego Morales Navarrete**, Pontificia Universidad Catolica de Chile, Ecuador
Co-authors: Luis Mauricio Castro, Moreno Bevilacqua, Christian Caamano Carrillo

Modelling spatial data is a challenging task in statistics. In many applications, the observed data can be modelled using Gaussian, skew-Gaussian, or even restricted random field models. However, in several fields, such as population genetics, epidemiology, and population dynamics, the data of interest are counts with excess of zeros in some cases, and therefore the mentioned models are not suitable for their analysis. Consequently, there is a need for spatial models that can adequately describe data coming from counting processes and handle the excess of zeros in data. Three approaches are commonly used to model this type of data, namely, GLMMs with Gaussian random field (GRF) effects, hierarchical models, and copula models. Unfortunately, these approaches do not explicitly characterize the random field like their q-dimensional distribution or correlation function. It is important to stress that GLMMs and hierarchical models induce a discontinuity in the path. Here, we propose a novel approach to efficiently and accurately model spatial count data with excess of zeros to deal with this. This approach is based on a random field characterization for count data with excess of zeros that inherit some well-known geometric properties from GFRs.

CO101 Room Aula E ECONOMETRICS METHODS FOR HIGH DIMENSIONAL DATA ANALYSIS Chair: Alessandra Amendola

C0158: **Estimating financial networks by realized interdependencies: A restricted vector autoregressive approach**
Presenter: **Massimiliano Caporin**, University of Padova, Italy
Co-authors: Deniz Erdemlioglu, Stefano Nasini

A network-based vector autoregressive approach is developed to uncover the interactions among financial assets by integrating multiple realized measures. Under a restricted parameter structure, our approach captures cross-sectional and time dependencies embedded in a large panel of assets. We propose a block coordinate descent procedure for the least square estimation and investigate its theoretical properties. Using U.S. data, we identify a large array of interdependencies with a limited computational effort. We also provide a new ranking for the systemically important financial institutions and carry out an impulse-response analysis to quantify the effects of adverse shocks on the financial system.

C0346: **Model structure identification in spatial econometrics**
Presenter: **Maria Lucia Parrella**, University of Salerno, Italy
Co-authors: Francesco Giordano, Marcella Niglio

Spatio-temporal data is often analyzed by means of *spatial econometric models*. In the last decade, several versions of these models have been proposed, each one based on specific assumptions and different properties of the estimators. We propose a strategy to identify the structure of a spatial econometric model for a given dataset, through a multiple testing procedure that allows choosing between a generalized version of the model and a nested version derived from the general one by imposing restrictions on the parameters. The proposal can be used to test the heterogeneity of the model, but also to test the presence of specific components, such as spatial effects, dynamic effects or external regressors. The theoretical properties of the testing procedure are derived in the high dimensional setup, where the number of spatial units grows to infinity with the sample size. A simulation study and an application to real data give empirical evidence of the testing procedure performance in presence of finite samples.

C0375: **Evolutionary correspondence analysis of the semantic dynamics of frames**
Presenter: **Giovanni Motta**, Texas A&M University, United States
Co-authors: Christian Baden

The aim is to introduce and implement a novel dimension-reduction method for high-dimensional time-varying contingency tables: the Evolutionary Correspondence Analysis (ECA). ECA offers new opportunities for the study of complex social phenomena, such as coevolving public debates: Its capacity to inductively extract time-varying latent variables from observed contents of evolving debates permits an analysis of meanings shared or exchanged between linked sub-discourses, such as linked national public spheres or distinct political camps and outlets within a shared public sphere. We illustrate the utility of our approach by studying how the Greek and German right-, center- and left-leaning news coverage of the European financial crisis evolved between its outbreak in 2009 until its institutional containment in 2012. Comparing the use of 525 unique concepts in six German and Greek outlets with different political leaning over an extended period of time, we identify two common factors accounting for those evolving meanings and analyze how the different sub-discourses influenced one another over time. We allow the factor-loadings to be time-varying, and fit to the latent factors a time-varying vector-auto-regressive model with a time-varying mean-vector. We provide identification conditions, asymptotic theory, and simulation results.

C0367: **Monitoring financial stress spillovers with high-frequency principal components**
Presenter: **Laura Garcia-Jorcano**, Universidad de Castilla-La Mancha, Spain
Co-authors: Massimiliano Caporin, Juan-Angel Jimenez-Martin

High-frequency principal components (HF PC) are used to extract information from stock prices to monitor and measure the existing level of stress in the financial system, which can be named the level of systemic stress, which is the amount of systemic risk that has already materialized. The empirical analysis using one-minute returns of stocks included in the Russel 3000 index from 2003 to 2021 shows that there exists a clear relationship between realized eigenvalues and systemic increases in financial stress. We also find that realized eigenvectors can trace the role of firms/sectors as potential sources of financial stress in different periods of time. Then, we measure the transmission of shocks from (to) the financial sector to (from) other non-financial sectors and the real economy. This provides a tool to analyze the spread of this financial instability that could affect the functioning of the financial system to the point where the real economy is seriously damaged. HF PC can be interpreted as a risk identification framework that allows policymakers and central banks to detect risks in good time and address potential threats to financial stability with the most appropriate policy tools.

CC156 Room Aula C HIGH-DIMENSIONAL STATISTICS I **Chair: Frank van der Meulen**

C0263: **High dimensional generalised penalised least squares**
Presenter: **Aikaterini Chrysikou**, Kings College, University of London, United Kingdom
Co-authors: Ilias Chronopoulos, George Kapetanios

Inference is developed for high dimensional linear models, with serially correlated errors. We examine the latter using the Lasso under the assumption of strong mixing in the covariates and error process, allowing for fatter tails in their distribution. While the Lasso estimator performs poorly under such circumstances, we estimate via penalised FGLS the parameters of interest and extend the asymptotic properties of the Lasso under more general conditions. Our theoretical results indicate that the non-asymptotic bounds for stationary dependent processes are sharper, while the rate of the Lasso under general conditions appears slower as $T, p \to \infty$. Further, we use the de-biased Lasso to perform inference on the parameters of interest. Using simulated data, we find that with the debiased generalised least squares estimator, our *t-tests* appear more powerful and correctly sized, while the true value of parameter is included in the 95% confidence interval with satisfying coverage rates at different levels of autocorrelation and parameter sparsity.

C0507: **The two-sample problem in high dimension: A ranking-based method**
Presenter: **Myrto Limnios**, University of Copenhagen, Denmark
Co-authors: Stephan Clemencon, Nicolas Vayatis

A general framework is proposed for testing the equality of two unknown probability distributions when considering two independent iid random samples, valued on a (same) measurable multivariate space. While there exists long-standing literature for the univariate setting, this problem remains a subject of research for both multivariate and nonparametric frameworks. Indeed, the increasing ability to collect large data of various structures, and possibly biased due to the collection process, for instance, has strongly defied classical modelings, particularly in applied fields such as biomedicine (clinical trials, genomics), marketing (AB testing). This method generalizes a particular class of permutation statistics known as two-sample linear rank statistics to multivariate spaces. By comparing the univariate image of the observations using a real-valued scoring function, a relation order is induced. The testing procedure is two-fold. 1) Maximization of the rank statistic: on the first half of each sample, we optimize a tailored version of the two-sample rank statistic over the class of scoring functions using ranking-based algorithms. 2) Two-sample homogeneity test: we perform the univariate rank test at fixed risk on the remaining observations, scored with the optimal scoring function of 1. Nonasymptotic theoretical guarantees are derived and numerical experiments modeling complex data structures compare and question both existing and present statistical tests.

C0606: **Debiased inference on heterogeneous quantile treatment effects with regression rank-scores**
Presenter: **Alexander Giessing**, University of Washington, United States
Co-authors: Jingshen Wang

Understanding treatment effect heterogeneity in observational studies is of great practical importance to many scientific fields. Quantile regression provides a natural framework for modeling such heterogeneity. We propose a new method for inference on heterogeneous quantile treatment effects in the presence of high-dimensional covariates. Our estimator combines an L1-penalized regression adjustment with a quantile-specific bias correction scheme based on quantile regression rank scores. We present a comprehensive study of the theoretical properties of this estimator, including weak convergence of the heterogeneous quantile treatment effect process to a Gaussian process. We illustrate the finite-sample performance of our approach through Monte Carlo experiments and an empirical example, dealing with the differential effect of statin usage for lowering low-density lipoprotein cholesterol levels for the Alzheimer's disease patients who participated in the UK Biobank study.

C0624: **Instrumental variable method in regularized regression with predictor measurement error**
Presenter: **Liqun Wang**, University of Manitoba, Canada
Co-authors: Lin Xue

Regularization methods are widely used in high-dimensional regression models and most methods are developed for the situation where all variables are correctly and precisely measured. However, in real-data analysis measurement error is common. We study the variable selection and estimation problems in linear and generalized linear models when some of the predictors are measured with error. We demonstrate how measurement error impacts the selection results and propose regularized instrumental variable methods to correct the measurement error effects. The proposed methods are consistent in selection and estimation and we derive their asymptotic distributions under general conditions. We also investigate the performances of the methods through Monte Carlo simulations and compare them with the naive method that ignores measurement error. Finally, the proposed method is applied to a real dataset.

CC233 Room Aula I COMPUTATIONAL STATISTICS AND APPLICATIONS **Chair: Roberto Di Mari**

C0687: **Time sensitive topic-based communities: The case of the vaccination debate in Italy**
Presenter: **Rebecca Graziani**, Bocconi University, Italy
Co-authors: Amelia Compagni

The aim is to reconstruct the debate that was developed in Italy around the compulsory vaccination of children of school age and culminated in 2017 with the emanation of a decree law by the Italian government. We analysed all public statements released about the issue to ANSA, the most important news agency in Italy. The corpus was assembled by retrieving all ANSA statements between January 1st 2015 and January 21st 2019, using the word vaccini (vaccines). We created a corpus of 3,225 statements, that we annotated and manipulated so to analyse sub-corpora based on the author, organization or date and as such to compare the statements produced by different subsets of actors or at different time periods. With the aim of identifying the main topics in the debate, we ran a topic analysis, with the Latent Dirichlet Allocation approach. The solution with fifteen topics ended out to be the best one in terms of coherence measures and interpretability. We restricted our attention to a selection of authors and implemented a network analysis on the time series of estimated topic weights by authors, so to identify time-sensitive topic-based communities. The analysis leads to the identification of three thematic communities.

C0689: **Assessing similarity among groups and global components in a dual STATIS multiple correspondence analysis**
Presenter: **Aida Eslami**, Laval University, Canada
Co-authors: Lauren Faye Toogood, Herve Abdi

In multivariate analysis, when variables are categorical the standard descriptive exploratory method is multiple correspondence analysis (MCA). MCA assumes that observations are independent and originate from a homogeneous population. However, the observations often comprise several groups known a priori (e.g., sex, ethnicity), a configuration known as multi-group data structure. In multi-group data, individuals from the same group are likely to be more similar to each other than to individuals from other groups. To take into account the group structure in MCA, we recently developed a new method called dual STATIS-MCA. This method lets us create both global and group components and loadings. In addition, to measure the similarity between these global and group components and loadings we used different approaches based on (1) the vector correlation (RV) coefficients, a multivariate generalization of the squared Pearson correlation coefficient, (2) Tuckers congruence coefficient, and (3) a method we previously developed. We illustrate this new procedure with a real case study.

C0686: **Multiple change point detection in functional sample via G-sum process**
Presenter: **Tadas Danielius**, Vilnius University, Lithuania
Co-authors: Alfredas Rackauskas

The G-cumsum process is defined and investigated in its theoretical aspects including asymptotic behavior. By choosing different sets G some tests for multiple change points detection in functional samples are proposed. The proposed testing procedures are applied to the real-world neurophysiological data and demonstrate how it can identify the existence of the multiple change points and localize them.

C0697: **Modeling pediatric hypertension occurrence: A case study**
Presenter: **Maria Filomena Teodoro**, IST-ID - Associacao do Instituto Superior Tecnico para a Investigacao e Desenvolvimento, Portugal

Pediatric hypertension (PH) causes severe risk factors and its prevention and obstruction. Diagnostic criteria for PH are the main reference to the normal distribution of arterial pressure (AP) in healthy children and are based on the fact that the PA pediatrician increases with aids and with a corporal mass. The objective is to characterize the profile of the AP in a pediatric population in order to educate and evaluate the prevalence of PH and AP normal-height and analyze the relationship between PH/AP normal-height and demographic characteristics. To finish the final stages of a study on PH prevalence at the national level promoted to the PH Group (Portuguese Pediatric Society) we considered the most recent data and followed previous work to estimate HP prevalence in the Portuguese population and to identify some associated risk factors. Statistical techniques like GLM or FA were considered.

CP205 Room Virtual Posters Room II POSTER SESSION II **Chair: Cristian Gatu**

C0321: **Analyzing big data on early literacy acquisition based on student interactions with digital reading supplement**
Presenter: **Yawen Ma**, Lancaster University, United Kingdom
Co-authors: Anastasia Ushakova, Kate Cain, Harrison Gamble, Jennifer Zoski

The focus is on the development of an exploratory analysis framework to be used on a large and complex dataset that captures the interactions of school children with a digital reading support supplement. The aim is to understand what contributes to early literacy acquisition using big data. The data arrives comes from student interactions with a research-based game environment, Amplify, which is composed of a variety of interactive games designed to support various reading skills. There is evidence that performance on assessments of various reading skills (e.g., morphological awareness, word reading) are interdependent and reciprocal, therefore, making which makes modelling the predictors of reading comprehension for beginner readers challenging. To characterize children's performance in various games we will evaluate how longitudinal models (e.g., latent growth curve) could be used. We will then use the results of the models to define if there are emerging clusters of behaviors. We can then look at the interdependence of game performance within those clusters using a simple graphical model structure. We will provide the foundation not only for a novel framework that can be applied to reading games big data but also will help to answer pressing research questions within literacy research (e.g., how different skills interplay dynamically in children's learning).

C0526: **Automatic predictor of diabetes mellitus, type i and gestational, using machine learning techniques**
Presenter: **Antonio Monleon-Getino**, Fundacio Bosh Gimpera, Spain
Co-authors: Ricardo Stalin Borja-Robalino, Karina Gibert, Gladys Robalino-Izurieta, Brigith Borja-Robalino, Jorge Buelvas-Muza, Raul Lopez-Torres, Carmen Serrano-Munuera

Currently, data mining presents a massive development and optimization of Data Mining devices and algorithms that identify complex patterns through the development of systems that learn autonomously. On the other hand, the incidence and prevalence of diabetes have increased in recent decades in all countries, as a consequence of the decrease in life expectancy and the increase in unhealthy habits. In Ecuador, diabetes has a prevalence of 4.7% in the population aged 10 to 59 years, while in Spain national surveys reflect a rate of about 8 out of 100 people. This research develops an automatic predictor of Diabetes, through Machine Learning techniques; becoming the first automatic predictive model at the national level for the prevention of Diabetes Mellitus, Type I and Gestational, trained with data from Hospitals in Ecuador and Spain. The choice of the model was based on the comparison of various techniques such as Logistic Regression, Linear Discriminant Analysis, Decision Trees, Naive Bayes, Support Vector Machine and Extreme Gradient Augmentation. The last model is found to be the most effective and efficient (83% accuracy) for the prediction at the level of Ecuador and Spain. The evaluation of models used the "EvaluaClas" library for R, published previously and that allows the standardization of performance metrics for machine and deep learning classifiers.

C0540: **How to measure relatedness between datasets during external validation of a multivariable prediction model**
Presenter: **Harald Heinzl**, Medical University of Vienna, Austria
Co-authors: Harbajan Chadha-Boreham, Martina Mittlboeck

The purpose of a multivariable prediction model (MPM) in clinical practice is the diagnosis or prognosis of a disease. Before routine use, the performance of the MPM has to be evaluated via internal and external validation. The main task of external validation is the assessment of the MPM's generalisability, an umbrella term for reproducibility and transportability. If the development and validation population are closely related, then an external validation study assesses reproducibility, otherwise, it assesses transportability. Hence, relatedness measures between the development and validation population have to be defined and appropriately estimated from the corresponding datasets. Since the development dataset will usually not be available during external validation, these measures have to be based on those basic summary statistics (e.g. percentages, means and standard deviations) that are commonly reported in medical research papers. Three proposals for such measures will be presented. They will be exemplified by an external validation study of the Framingham steatosis index.

C0553: **Bayesian approach for modelling rna transcription**

Presenter: **Elena Sabbioni**, Politecnico di Torino, Italy

Co-authors: Gianluca Mastrantonio, Enrico Bibbona, Guido Sanguinetti

Gene expression is regulated through the fundamental process of transcription, splicing and degradation, which can be modelled as an ODE system, whose parameters need to be estimated from experimental data collected by single-cell RNA sequencing. By this technique, biologists can take only a single snapshot of the cellular states: they obtain the counts of unspliced and spliced mRNA molecules, for each gene and for each gene altogether at the moment of the sequencing, which actually corresponds to different levels of maturity in the evolution of the different cells, and then the cells are destroyed. The aim is to reconsider part of the methods currently used to estimate the parameters of the model and to describe the evolution of some cells over different cell types, exploiting the level of expression of their genes and the concept of RNA-velocity. We reformulate it in a way that is mathematically better founded, using Bayesian statistics. We discuss the advantages of this approach in terms of the quality and the interpretability of the results.

C0560: **Using correlated resampling to improve variable selection for linear and generalized linear models**

Presenter: **Myriam Maumy**, IRMA/Universite of Technology of Troyes, France

Co-authors: Frederic Bertrand

Technological innovations make it possible to measure large amounts of data in a single observation. Hence, problems in which the number of variables is larger than the number of observations have become common. As reviewed almost twenty years ago, such situations arise in many fields from fundamental sciences to social science, and variable selection is required to tackle these issues. Moreover, in such studies, the correlation between variables is often very strong, and variable selection methods often fail to make the distinction between the informative variables and those which are not. As a consequence, variable selection has become one of the critical challenges in statistics and many methods have already been proposed in the literature. If the number of variables far exceeds the number of observations or if the variables are highly correlated, performances of variable selection methods are generally limited in recall and precision. We propose a general algorithm that enhances model selection in correlated variables dataset. We use the correlation structure to select reliable variables in parsimonious or non-parsimonious linear regression or generalized linear regression problems. Thanks to correlated resampling techniques, it is possible to improve the performance of many common existing models -glmnet, lasso, spls, - as demonstrated on both simulated and real datasets using a comprehensive simulation benchmark.

C0562: **Bootstrap-based hyper parameter tuning for sparse partial least squares regular or generalized regressions**

Presenter: **Frederic Bertrand**, IRMA/Universite de technologie de Troyes, France

Co-authors: Myriam Maumy

Methods based on partial least squares (PLS) regression, which has recently gained much attention in the analysis of high-dimensional genomic datasets, have been developed since the early 2000s for performing variable selection. Most of these techniques rely on tuning parameters that are often determined by cross-validation (CV) based methods, which raises essential stability issues. To overcome this, we have developed a new dynamic bootstrap-based method for significant predictor selection, suitable for both PLS regression and its incorporation into generalized linear models (GPLS). It relies on establishing bootstrap confidence intervals, which allows testing of the significance of predictors at preset type I risk, and avoids CV. We have also developed adapted versions of sparse PLS (SPLS) and sparse GPLS regression (SGPLS), using a recently introduced non-parametric bootstrap-based technique to determine the numbers of components. We compare their variable selection reliability and stability concerning tuning parameters determination and their predictive ability, using simulated data for PLS and real microarray gene expression data for PLS-logistic classification. We observe that our new dynamic bootstrap-based method has the property of best separating random noise in y from the relevant information with respect to other methods, leading to better accuracy and predictive abilities, especially for non-negligible noise levels.

C0579: **A nearest neighbours Gaussian Process model for time-frequency data: An application in bio-acoustic analysis**

Presenter: **Hiu Ching Yip**, Politecnico Di Torino, Italy

Co-authors: Gianluca Mastrantonio, Enrico Bibbona, Marco Gamba, Daria Valente

In comparative bio-acoustic studies, one area of interest is to understand the acoustic structures of non-human primates in order to provide insights into the evolution of the communication mechanism of our closest relatives. The most common practices are feature engineering methods, which involve selecting a set of basis-features for quantitative comparison. The identification of meaningful features in the vocal repertoire relies on biologists to observe and interpret the behavioural contexts in which the animals emit the signals. These interpretations are costly to acquire, inaccurate due to human subjectivity and difficult to generalize for cross-species comparison. Furthermore, feature selection always treats the time-frequency bins of spectrograms as independent features or extracts common statistics from waveforms. This ignores the time-varying effect of observed vocalizations on the latent acoustic structure as well as the periodic nature of time-frequency data. The aim is to propose a Nearest Neighbour Gaussian Process model that accounts for the time varying components as well as the circular nature of time-frequency for latent spectral structure inference from bio-acoustic data. The dataset that will be available for model implementation are vocal signals of lemurs that were recorded in Madagascar.

C0662: **Supervised and ANN classification to physical activities performed by pregnant women recorded via accelerometer data**

Presenter: **Gleici Perdona**, USP, Brazil

Co-authors: Rafael Biagioni Fazio, Christoph Michael Mitschka

Physical activity, even if just of light intensity, is helpful to mother and fetal health during pregnancy and after delivery. It is critical to analyze the amount of physical activity among pregnant women from various socioeconomic backgrounds and lifestyles, to better understand the elements that influence their physical activity habits. This research aims to define and classify physical activities undertaken by 150 pregnant women in the Ribeiro Preto city in Brazil at the Unified Health System (SUS), based on data generated using accelerometers and an application to record the performed physical activities. To assess the need for physical activity, a virtual assistant is being developed in the project https://eva.fmrp.usp.br/ - EVA, which aims to follow pregnant women during this period and analyze their need for physical activity and based on the results, make recommendations. Consequently, caring for the health of the pregnant woman. For this purpose, data cleaning and processing techniques were applied and then, for supervised classification, were considered LightGBM (tree-based gradient boosting) and artificial neural networks of the type of Long short-term memory (LSTM). Among the conclusions, training using a 30-second period is pointed out as the approach with the best accuracy metrics. Some weaknesses of this approach were also identified, and possible improvements were derived.

C0702: **Advanced HMMs for physiological time series**

Presenter: **Kristian Romano**, University of Warwick, United Kingdom

Co-authors: Sida Chen, Barbel Finkenstadt, Francis Levi

Wearable devices allow for non-invasive telemonitoring of patients. Physiological data, such as physical activity and body temperature, can be collected in the daily living environment of the patient without the need for hospitalization. The devices can be used to monitor the Circadian Timing System (CTS), which regulates many critical cellular processes, such as the cell cycle and metabolism, and to detect adverse events which pose a risk to health. The appraisal of the CTS in real time could lead to chronopharmacological strategies and personalized medicine. To model the physiological data arising from wearable sensors we developed non-homogeneous hidden Markov and semi-Markov models with nonparametric emission distributions. To accommodate for the oscillating nature of the CTS, the transition probabilities are driven by a circadian oscillator. We will illustrate the novel models, estimation algorithms and some (preliminary) results for simulations.

Thursday 25.08.2022 16:15 - 17:45 Parallel Session L – COMPSTAT2022

CV196 Room Aula B MACHINE LEARNING (VIRTUAL) Chair: Alejandro Murua

C0477: **The SgenoLasso for gene mapping and genomic prediction**
Presenter: **Charles-Elie Rabier**, Montpellier University, France
Co-authors: Celine Delmas

The focus is on the problem of detecting Quantitative Trait Loci, so-called QTL (genes influencing a quantitative trait which can be measured) on a given chromosome $[0,T]$. We assume a linear model on the quantitative trait $Y_i = \mu + \sum_{s=1}^{m} X_i(t_s^\star)q_s + \sigma\varepsilon_i$ where μ is the global mean, $X_i(\cdot)$ the genome information, ε_i a Gaussian white noise, σ^2 the environmental variance, m the number of QTL, q_s and $t_s^\star$ denote respectively the QTL effect and the location of the sth QTL. The quantitative trait Y_i is measured on all the individuals i whereas the genome information X_i is available only on extreme individuals and only at fixed locations. First, we derive theoretical properties of the score test process and likelihood ratio test process along the chromosome under the null hypothesis of no QTL on $[0,T]$ and under the alternative hypothesis that there exits m QTL on the chromosome. We deduce a new method, called SgenoLasso, to estimate the number of QTL, their locations, and their effects. This method will also be used for genomic prediction. It will be compared to classical methods (Lasso, Group Lasso, Elastic Net, RaLasso, Bayesian Lasso) on simulated data and applied to real data.

C0512: **Large-scale entropy regularized optimal transport independence criterion**
Presenter: **Lang Liu**, University of Washington, United States
Co-authors: Soumik Pal, Zaid Harchaoui

An independence criterion is introduced based on entropy regularized optimal transport (EOT). Its empirical estimator involves solving an EOT problem between two discrete distributions whose support size scales quadratically in the sample size n. This is computationally challenging since a naïve application of the popular Sinkhorn algorithm requires $O(n^4)$ time and space. For large-scale problems, we design a Tensor Sinkhorn algorithm equipped with a random feature type approximation, reducing the time complexity and space complexity to $O(n^2)$. We also offer a differentiable program implementation for deep learning applications, which allows one to run the reverse mode automatic differentiation through statistical quantities based on our criterion. We present experimental results on existing benchmarks for independence testing, illustrating the interest of the proposed criterion to capture both linear and nonlinear dependencies in synthetic data and real data.

C0571: **Actual events vs. perceived reporting: Modeling firm performance under environmental uncertainty using machine learning**
Presenter: **Minh Nguyen**, University of Hawaii at Manoa, United States

Not all companies respond the same to natural disaster events. The aim is to investigate two ways that natural disasters affect firm performance: actual events vs. perceived reporting. We consider the billion-dollar weather and climate disasters in the United States as the actual events and the number of words related to natural disasters in the Management Discussion and Analysis section in Form 10-Ks filing by the U.S. public companies as the perceived reporting. The aim is also to compare the performances of classification and regression trees (CART) and neural networks with linear regression in predicting the performance of U.S. public companies under environmental uncertainty. The results show that both actual events and perceived reporting of natural disasters this year negatively affect the return on assets in the next year. Also, the actual natural disasters this year negatively affect the next year's sales growth. An important result of neural networks is that deeper networks do not ensure improving the predictive accuracy in predicting firm performance. Comparing CART, neural networks, and linear regression models, we find that CART and neural networks outperform linear regression models in predicting firm performance. This result is robust to any given firm performance criteria, split ratios, and prediction errors.

CI107 Room Aula F CAUSALITY AND DISTRIBUTIONAL ROBUSTNESS (VIRTUAL) Chair: Armeen Taeb

C0288: **Calibrated inference: Statistical inference that accounts for both sampling uncertainty and distributional uncertainty**
Presenter: **Dominik Rothenhaeusler**, Stanford University, United States
Co-authors: Yujin Jeong

During data analysis, analysts often have to make seemingly arbitrary decisions. For example during data pre-processing, there are a variety of options for dealing with outliers or inferring missing data. Similarly, many specifications and methods can be reasonable to address a certain domain question. This may be seen as a hindrance to reliable inference since conclusions can change depending on the analyst's choices. We argue that this situation is an opportunity to construct confidence intervals that account not only for sampling uncertainty but also for some type of distributional uncertainty. Distributional uncertainty is closely related to other issues in data analysis, ranging from dependence between observations to selection bias and confounding. We demonstrate the utility of the approach on simulated and real-world data.

C0427: **Distribution generalization with instrumental variables**
Presenter: **Niklas Pfister**, University of Copenhagen, Denmark
Co-authors: Jonas Peters, Leonard Henckel, Sorawit Saengkyongam, Rune Christiansen, Sebastian Engelke, Martin Jakobsen, Nicola Gnecco

Causal models can provide good predictions even under distributional shifts. This observation has led to the development of various methods that use causal learning to improve the generalization performance of predictive models. We consider this type of approach for instrumental variable (IV) models. IV allows us to identify a causal function between covariates X and a response Y, even in the presence of unobserved confounding. In many practical prediction settings, the causal function is however not fully identifiable. We consider two approaches for dealing with this under-identified setting: (1) By adding a sparsity constraint and (2) by introducing the invariant most predictive (IMP) model, which deals with the under-identifiability by selecting the most predictive model among all feasible IV solutions. Furthermore, we analyze to which types of distributional shifts these models generalize.

C0535: **Causal structure learning with unknown interventions**
Presenter: **Armeen Taeb**, ETH Zurich, Switzerland

With observational data alone, causal inference is a challenging problem. The task becomes easier when having access to data collected from perturbations of the underlying system, even when the nature of these is unknown. We will describe a body of work that makes algorithmic and theoretical advances for identifying plausible causal mechanisms from such perturbation data. Specifically, in the context of Gaussian linear structural equation models, we first characterize the interventional equivalence class of DAGs. We then leverage these results to study high-dimensional consistency guarantees of a thresholded l_0 penalized maximum likelihood estimator for learning said class. Finally, we describe extensions to settings with latent variables.

CO138 Room Aula G HEAVY-TAILED DISTRIBUTIONS FOR FINANCIAL MODELING Chair: Marco Bee

C0250: **Automatic threshold selection for extreme value regression models**
Presenter: **Julien Hambuckers**, University of Liege, Belgium
Co-authors: Marie Kratz, Antoine Usseglio-Carleve

In finance, extreme value regression (EVR) has become a standard tool in the econometrician's toolbox to estimate and characterize risk measures in changing economic conditions. However, in this regression context, the threshold choice is a non-trivial task since it should also depend on the covariates and can have important consequences on the final estimates. We investigate this under-discussed issue and propose an efficient and robust solution to automatically estimate these thresholds with the help of the distributional regression machinery. We illustrate its properties through several simulation studies. The method is later applied to estimate hedge funds' tail risks, accounting for their heterogeneous investment strategies and the time-varying characteristics of financial markets.

C0259: **Modeling panels of extremes**
Presenter: **Luca Trapin**, University of Bologna, Italy
Co-authors: Debbie Dupuis, Sebastian Engelke

Extreme value applications commonly employ regression techniques to capture cross-sectional heterogeneity or non-stationarity in the data. Estimation of the parameters of an extreme value regression model is notoriously challenging due to the small number of observations that are usually available in applications. When repeated extreme measurements are collected on the same individuals, i.e., a panel of extremes is available, pooling the observations in groups can improve the statistical inference. We study three data sets related to risk assessment in finance, climate science, and hydrology. In all three cases, the problem can be formulated as an extreme value panel regression model with a latent group structure and group-specific parameters. We propose a new algorithm that jointly assigns the individuals to the latent groups and estimates the parameters of the regression model inside each group. Our method efficiently recovers the underlying group structure without prior information, and for the three data sets it provides improved return level estimates and helps answer important domain-specific questions.

C0313: **Robust score-driven filters and smoothers**
Presenter: **Debbie Dupuis**, HEC Montreal, Canada
Co-authors: Luca Trapin

Score-driven models are quickly gaining attention as simple methods to obtain fast and accurate approximate filters and smoothers for the latent states of state-space models. A class of robust score-driven filters and smoothers able to reduce the size of the approximation error is developed. A large simulation study suggests that our robust approach can provide large efficiency gains in terms of mean squared and absolute error compared to the standard score-driven filters and smoothers in several non-linear state-space examples. Two financial applications illustrate the benefits of robustification on real data.

C0347: **Forecasting in GARCH models with polynomially modified innovations**
Presenter: **Gianmarco Vacca**, Universita Cattolica del Sacro Cuore, Italy
Co-authors: Maria Grazia Zoia, Luca Bagnato

Orthogonal polynomials can be used to modify the moments of the distribution of a random variable. Polynomially adjusted distributions are employed to model the skewness and kurtosis of the conditional distributions of GARCH models. To flexibly capture the skewness and kurtosis of data, the distributions of the innovations that are polynomially reshaped include, besides the Gaussian, also leptokurtic laws such as the logistic and the hyperbolic secant. Modeling GARCH innovations with polynomially adjusted distributions can effectively improve the precision of the forecasts. This strategy is analyzed in GARCH models with different specifications for the conditional variance, such as the APARCH, the EGARCH, the Realized GARCH, and APARCH with time-varying skewness and kurtosis. An empirical application on different types of asset returns shows the good performance of these models in providing accurate forecasts according to several criteria based on density forecasting, downside risk, and volatility prediction.

CO027 **Room Aula C** **SURVEY SAMPLING** **Chair: Alina Matei**

C0329: **A multi-spreading algorithm to account for spatial and strata heterogeneity**
Presenter: **Maria Michela Dickson**, University of Trento, Italy
Co-authors: Yves Tille, Giuseppe Espa, Flavio Santi, Diego Giuliani

Spatial designs producing spatially spread samples permit estimate precision to be improved when the variable of interest exhibits some form of spatial heterogeneity. However, if the variable of interests exhibits heterogeneity also with respect to some partition of the target population, a global spread of the sample over the reference space may not result in efficient estimates. On the other hand, the same problem may arise if sampled units are spatially spread only within each population stratum. We propose a sampling algorithm which considers both sources of heterogeneity and produces samples which are spatially spread both globally and within each stratum, according to a tuning parameter $\alpha \in [0,1]$, that globally spreads the sample when $\alpha = 1$, and spreads each sub-sample independently when $\alpha = 0$. Formal analysis and Monte Carlo simulations showed that the proposed algorithm can effectively exploit the trade-off between the global and within-stratum spread of the sample, and produce efficiency gains when the parameter is properly set.

C0516: **Spbsampling: An R package for spatially balanced sampling**
Presenter: **Francesco Pantalone**, University of Southampton, United Kingdom
Co-authors: Roberto Benedetti, Federica Piersimoni

In environmental, geological, biological, and agricultural surveys, among many others, usually, the main feature of the population of interest is to be geo-referenced. In these situations, we can expect that units closer to each other provide less information about a target of inference than units farther apart, as outlined in Tobler's first law of Geography. Therefore, it would be beneficial to the efficiency of the final estimates to consider the spatial dependence. Since traditional sampling designs generally do not take into account the spatial features of the population, several spatially balanced sampling designs have been introduced in the literature, which select samples well spread over the population of interest, or spatially balanced samples. We introduce the R package Spbsampling, which implements some of the designs recently introduced. We focus on sampling designs that achieve spatially balanced samples by means of an MCMC algorithm and the use of a summary index of a distance matrix. This allows for wider applicability, as a distance matrix can be defined for units according to variables different from geographical coordinates.

C0651: **Model-assisted estimators in surveys with nonresponse**
Presenter: **Caren Hasler**, University of Neuchatel, Switzerland
Co-authors: Esther Eustache

In the presence of auxiliary information, model-assisted estimators use a working model that links the variable of interest and the auxiliary variables in order to improve the Horvitz-Thompson estimator. The resulting estimators are asymptotically designed unbiased and asymptotically more efficient than the Horvitz-Thompson estimator under some regularity conditions and for a wide range of working models. We adapt model-assisted total estimators to missing at random data building on the idea of nonresponse weighting adjustment. We see nonresponse as a second phase of the survey and reweight the units in model-assisted estimators using the inverse of estimated response probabilities in order to compensate for the nonrespondents. We develop the asymptotic properties of our proposed estimators and discuss the calibration of the weights of these estimators. We provide formulae for asymptotic variance and variance estimators. We conduct a simulation study to describe the behavior of the proposed estimators.

C0393: **Solutions inspired by survey sampling theory to build effective clinical trials**
Presenter: **Yves Tille**, University of Neuchatel, Switzerland

The organization of a design of experiments, for example for the realization of a clinical trial, is crucial. It is often desirable to balance designs so that the means of the covariates are approximately the same in the test and control groups. In survey sampling theory, balanced sampling and calibration are two techniques that improve the precision of estimates. We show the links between the two areas. We begin by assessing the gain in precision between a balanced design and a simple random sampling for the least squares estimators and the estimator by differences. We compare rerandomization techniques and the cube method in order to balance the design. We propose a new method, particularly efficient, which combines the cube method with multivariate matching. A set of simulations is carried out in order to evaluate the different methods. The interest in the calibration is shown even if the design is almost balanced. It is thus shown that tools used by survey statisticians can be useful for experimental designs and clinical trials.

CO025 Room Aula D NEW INSIGHTS IN ROBUST METHODS OF INFERENCE **Chair: Laura Ventura**

C0366: **Composite Tsallis score: A tool for robust inference**
Presenter: **Valentina Mameli**, University of Udine, Italy
Co-authors: Monica Musio, Erlis Ruli, Laura Ventura

Classical likelihood inference can be difficult to perform both when the full likelihood is too complex or even impossible to specify and when robustness with respect to data or to model misspecification is required. In these situations, in order to perform inference, it may be useful to consider suitable pseudolikelihoods. These would include, for example, composite likelihoods that are constructed by composing low-dimensional likelihood objects and forming a subset of a more general class of methods based on proper scoring rules. Proper scoring rules, other than the logarithmic score, can be used as an alternative to the full likelihood when the interest is in increasing robustness or simplifying computations. Examples of particular interest include the Tsallis score which in general gives robust procedures. To address both complex models and model misspecification, we propose to resort to a composite robust scoring rule. In particular, we focus on the pairwise Tsallis score, obtained as a weighted sum of Tsallis scores for marginal or conditional bivariate events.

C0450: **Filters based on statistical data depths for robust multivariate inference**
Presenter: **Claudio Agostinelli**, University of Trento, Italy
Co-authors: Giovanni Saraceno

In the classical contamination models, such as the Huber-Tukey contamination model (Case-wise Contamination), observations are considered as the units to be identified as outliers or not. This model is very useful when the number of considered variables is moderately small. It has been shown that the limits of this approach for a larger number of variables and introduced the Independent contamination model (Cell-wise Contamination) where the cells are the units to be identified as outliers or not. One approach to deal, at the same time, with both types of contamination is to filter out the contaminated cells from the data set and then apply a robust procedure able to handle case-wise outliers and missing values. Here we develop a general framework to build filters in any dimension based on statistical data depth functions. We show that previous approaches to construct filters are special cases. We discuss the main theoretical properties of our method and illustrate its performance by Monte Carlo simulation and examples.

C0455: **Resistant inference for complex and large models**
Presenter: **Maria-Pia Victoria-Feser**, University of Geneva, Switzerland
Co-authors: Yuming Zhang

At this moment in time, data not only come in huge quantities, but they also come with features that are not desirable. These include outliers (that are typically hard to detect), missing data, selection bias, measurement errors, and so on. Although there exists an abundance of (easily accessible) statistical and computational methods, these methods tend to address these features separately. Among others, one potential reason for this situation is that accounting for these data features simultaneously can encounter enormous hindrances in the computational aspects. We propose an alternative robustness framework that allows the construction of resistant estimators and associated inferential procedures that have desirable finite sample properties and are computationally tractable with complex and large models. This framework furthermore allows including additional data features such as informative missingness, selection bias and/or measurement errors, with a small additional price in terms of computational costs. More specifically, we employ a two-step approach consisting of considering first a naive estimator that is easy to compute, and derive from it, using a simulation-based approach, a final estimator with desirable asymptotic and finite sample properties. We apply the proposed methodology to GLM and MLM with censored or misclassified data and with outliers.

C0458: **E is the new P: Optional continuation and evidence**
Presenter: **Peter Grunwald**, CWI and Leiden University, Netherlands

How much evidence do the data give us about one hypothesis versus another? The standard way to measure evidence is still the p-value, despite a myriad of problems surrounding it. One central problem is its inability to deal with optional stopping, combining evidence of separate studies and its dependence on unknowable counterfactuals. We present the E-value, a recently popularized notion of evidence which overcomes these issues. If the null hypothesis is simple and there is an alternative, the E-value coincides with the Bayes factor, the notion of evidence preferred by Bayesians. But if the null is composite or nonparametric, or an alternative cannot be explicitly formulated, E-values and Bayes factors become distinct. Unlike the Bayes factor, E-values allow for tests with strict Type-I error control. They are also the basic building blocks of anytime-valid confidence intervals that remain valid under optional stopping.

CO136 Room Aula I RECENT DEVELOPMENTS IN HIGH-DIMENSIONAL STATISTICS **Chair: Shubhadeep Chakraborty**

C0339: **A multivariate permutation test for the analysis of paired samples: the mixed data scenario**
Presenter: **Riccardo Ceccato**, Universita degli Studi di Padova, Italy
Co-authors: Rosa Arboretti, Elena Barzizza, Nicolo Biasetton, Marta Disegna, Luca Pegoraro, Luigi Salmaso

The comparison of two populations can be quite challenging under multivariate scenarios, especially when the number of variables is much larger than the sample size. The Nonparametric Combination (NPC) methodology represents a suitable, flexible and quite powerful solution to such problems, allowing us to implicitly take into account existing correlations in a multivariate outcome. Additionally, the NPC can be adopted to deal with several complex scenarios involving multiple data types. We focus on a particularly challenging task, in which paired samples need to be compared and both numeric and ordinal variables are available, and propose an appropriate NPC-based solution. A simulation study is conducted to evaluate the performance of our proposal.

C0486: **Nonparametric sequential change-point detection in high dimensions**
Presenter: **Shubhadeep Chakraborty**, University of Washington, Seattle, USA, United States

Sequential change-point detection is a classical problem in statistics where the goal is to detect a change in the data generating mechanism as soon as possible after it occurs. We propose two nonparametric algorithms and stopping rules to sequentially detect general distributional changes along a high-dimensional data stream, rather than detecting changes only in the mean or in the covariance structure. The first algorithm adopts a sliding window-based approach that performs a sequence of two-sample tests for homogeneity between a post-change sample and a reference pool for every newly recorded observation. A theoretical approximation of the average run length is rigorously derived, which enables us to easily obtain the value of the threshold for the stopping rule, thus achieving control on the false alarm rate. We also establish an explicit upper bound for the expected detection delay and illustrate the impact of certain key factors on the expected detection delay. In the second algorithm, we construct a sequence of thresholds for the sequential two-sample tests using generalized alpha investing rules, controlling the false discovery rate. Numerical studies illustrate the superior performance of our algorithms over other state-of-the-art methods.

C0362: **DisCo P-ad: Distance Correlation-based P-value Adjustment boosts multiple-testing corrections in metabolomics analyses**
Presenter: **Debmalya Nandy**, Colorado School of Public Health, United States
Co-authors: Debashis Ghosh, Katerina Kechris

High-throughput data, often encountered in -omics sciences (e.g., genomics, metabolomics), contain measurements on several hundred or thousands of variables. In tests of association of these predictors with a clinical outcome of interest, multiple-testing corrections mitigate the number of false and truly missed discoveries. Many corrections involve first estimating the effective number of tests (number of statistically independent predictors among all original ones) for a subsequent Bonferroni-type adjustment to obtain the point-wise significance level, corresponding to a preset overall type-I error rate. Such practice is commonplace in Genome-Wide Association Studies (GWAS) but is also relevant to Metabolome-Wide Association Studies (MWAS). For MWAS, we consider procedures for p-value adjustments in GWAS along with one specifically designed for MWAS. While most are based on eigen-analysis of the Pearson's correlation matrix of the predictors, we propose using the Distance Correlation instead in the eigen-analysis for P-value adjustment (DisCo P-ad). Our extensive simulation study, based on real metabolomics datasets, demonstrates superior performance of DisCo for varying sample sizes, nature of the response (continuous/binary), and groupings of the metabolites. In summary, our study introduces a new method for enhancing multiple testing corrections in MWAS.

C0267: **Selective inference for k-means clustering**
Presenter: **Yiqun Chen**, University of Washington, Seattle, United States
Co-authors: Daniela Witten

The problem of testing for a difference in means between clusters of observations identified via k-means clustering is considered. In this setting, classical hypothesis tests lead to an inflated Type I error rate. To overcome this problem, we take a selective inference approach. We propose a finite-sample p-value that controls the selective Type I error to test the difference in means between a pair of clusters obtained using k-means clustering, and show that it can be efficiently computed. We apply our proposal in simulations and on hand-written digits data and single-cell RNA-sequencing data.

CO055 Room Aula Q BIOSTATISTICS AND BIOCOMPUTING — Chair: Yisheng Li

C0438: **A multiple imputation-based sensitivity analysis approach for regression analysis with a MNAR covariate**
Presenter: **Chiu-Hsieh Hsu**, University of Arizona, United States
Co-authors: Yulei He, Chengcheng Hu, Wei Zhou

Missing covariate problems are common in biomedical studies. If there is a suspicion of missing not at random, researchers often perform sensitivity analysis to evaluate the impact of various missingness mechanisms. Under the selection modeling framework, we propose a sensitivity analysis approach with a standardized sensitivity parameter using a nonparametric multiple imputation strategy. The proposed approach requires fitting two working models for deriving two predictive scores and specifying the correlation coefficient between missing covariate values and selection probabilities. For each missing covariate observation, the two predictive scores are used to select the nearest neighborhood and the correlation coefficient is used to define an imputing set based on the selected neighborhood. The proposed approach is expected to be more robust against mis-specifications of the selection model and the sensitivity parameter since the selection model is only used to induce missing not at random and the sensitivity parameter is only used to define imputing sets. For illustration, the proposed approach is applied to evaluate the relationship between post-operative outcomes and incomplete pre-operative Hemoglobin A1c levels for surgical high-grade carotid artery stenosisa patients. A simulation study is conducted to evaluate the performance of the proposed approach.

C0616: **A uniform shrinkage prior in spatio-temporal Poisson models for count data**
Presenter: **Yisheng Li**, The University of Texas MD Anderson Cancer Center, United States

Default Bayesian inference is considered in a Poisson generalized linear mixed model for spatio-temporal data. Normal random effects are used to model the within-area correlation over time and spatial effects represented with a proper conditional autoregressive model are used to model the between-area correlations. We develop a uniform shrinkage prior for the variance components of the spatiotemporal random effects. We prove that the proposed USP is proper, and the resulting posterior is proper under the proposed USP, an independent flat prior for each fixed effect, and a uniform prior for a spatial parameter, under suitable conditions. Posterior simulation is implemented and inference is made using the OpenBUGS, R2OpenBUGS and RStan software packages. We illustrate the proposed method by applying it to a leptospirosis count dataset with observations from 17 northern provinces of Thailand across four quarters in 2011 to construct the disease maps. According to the deviance information criterion, the proposed USP for the variance components of the spatiotemporal effects yields better performance than the conventional inverse gamma priors. A simulation study suggests that the estimated fixed-effect parameters are accurate, based on a relative bias criterion.

C0644: **Posterior predictive design for phase I clinical trials**
Presenter: **Shouhao Zhou**, Penn State University, United States

Model-assisted designs are cutting-edge adaptive designs to find the maximum tolerated dose (MTD) in phase I clinical trials. They enjoy superior performance compared to more complicated, model-based adaptive designs, but with their decision rule pretabulated, they can be implemented as simply as the conventional algorithmic designs. We propose the posterior predictive (PoP) design, a novel model-assisted design based on Bayesian interval hypothesis testing for dose escalation and de-escalation. The work moves beyond the previous model-assisted designs by theoretically achieving consistency in selecting the true MTD and global optimality in dose transition. The simulation studies demonstrate that the PoP design can achieve significant improvement in operating characteristics to identify the MTD, thereby providing a useful upgrade to the prevalent model-assisted designs.

C0665: **Bayesian nonparametric analysis for the detection of spikes in noisy calcium imaging data**
Presenter: **Michele Guindani**, University of California, Irvine, United States

Recent advancements in miniaturized fluorescence microscopy have made it possible to investigate neuronal responses to external stimuli in awake behaving animals through the analysis of intra-cellular calcium signals. An ongoing challenge is deconvoluting the temporal signals to extract the spike trains from the noisy calcium signals' time-series. We propose a nested Bayesian finite mixture specification that allows for the estimation of spiking activity and, simultaneously, reconstructing the distributions of the calcium transient spikes' amplitudes under different experimental conditions. The proposed model leverages two nested layers of random discrete mixture priors to borrow information between experiments and discover similarities in the distributional patterns of neuronal responses to different stimuli. Furthermore, the spikes' intensity values are also clustered within and between experimental conditions to determine the existence of common (recurring) response amplitudes. Simulation studies and the analysis of a data set from the Allen Brain Observatory show the effectiveness of the method in clustering and detecting neuronal activities.

CO119 Room Aula E ADVANCES IN K-MEANS AND CLUSTERING ENSEMBLE METHODS — Chair: Roberta Pappada

C0353: **Model ensemble in density-based clustering**
Presenter: **Alessandro Casa**, Free University of Bozen-Bolzano, Italy

Co-authors: Luca Scrucca, Giovanna Menardi

Model-based clustering represents a widely known approach when searching for groups in the data. Finite mixture models are adopted to describe the data generative mechanism, and partitions are obtained by drawing a correspondence between components and groups. Operationally several different models, with different parameterizations and number of components, are estimated and the best one among them is chosen by means of an information criterion. Nonetheless, considering a single model to cluster the data, this strategy may be sub-optimal since throwing away all the fitted models except for the best one could lead to a potentially harmful loss of information. In order to overcome this issue, an ensemble clustering approach is proposed, circumventing the single best model paradigm, thus potentially improving the stability and the robustness of the partitions. A new density estimator, defined as a convex linear combination of the density estimates in the ensemble, is introduced and exploited for group assignment. Finally, since the correspondence between mixture components and clusters is lost in the process, we define partitions by borrowing the modal, or nonparametric, formulation of the clustering problem, where groups are associated with high-density regions of the density.

C0474: **Pivotal consensus clustering through the `pivmet` R package**

Presenter: **Leonardo Egidi**, University of Trieste, Italy

Co-authors: Roberta Pappada, Francesco Pauli, Nicola Torelli

Identifying the prototypes of a group, i.e. the elements of a dataset representing different groups of data points, is relevant to the tasks of clustering, classification and mixture modeling. The R package pivmet provides functions to perform consensus clustering based on pivotal units, which may allow the improvement of classical techniques such as the standard k-means algorithm via careful seeding; moreover, the package is flexibly programmed to support applications to real and simulated datasets. Finally, some preliminary procedures aimed at identifying the number of clusters based on a co-association matrix will be illustrated.

C0475: **HC-fused: A versatile R-package for multi-omics hierarchical ensemble clustering**

Presenter: **Bastian Pfeifer**, Medical University of Graz, Austria

Co-authors: Michael Georg Schimek

An important application of multi-omics clustering methods is the stratification of patients into sub-groups of similar molecular characteristics. In recent years many advancements in such methods have been developed, that provide a deeper understanding of cancer progression and may facilitate oncological treatment. However, due to the high diversity of cancer-related data, a single method may not perform sufficiently well in different scenarios. We offer a versatile framework for multi-omics hierarchical ensemble clustering, implemented within the R-package HC-fused. HC-fused allows for building hierarchical clustering ensembles suitable for the available data and research goals. In addition, a data fusion approach, developed by us, combines the clustering results from different ensemble methods and/or omics data sets, and at the same time allows the user to track the individual contribution of each single-omic and/or method to the data fusion process. Survival analyses for data from The Cancer Genome Atlas (TCGA) indicate that our proposed ensembles provide more robust, and thus more reliable results than state-of-the-art approaches. The mentioned methodology is implemented in the R-package HC-fused freely available from GitHub (https://github.com/pievos101/HC-fused).

C0484: ***K*-means algorithm with positive and negative equivalence constraints**

Presenter: **Igor Melnykov**, University of Minnesota Duluth, United States

A modification of the K-means algorithm is considered that accommodates two types of hard constraints often encountered in semi-supervised clustering. A positive equivalence constraint requires that the data points bound by such a constraint are placed in the same class in the clustering solution. At the same time, a negative constraint specifies which points must be separated from each other and included in different classes. Although it is common to check for any constraints in the form of an add-on to the basic K-means algorithm, its objective function is usually left unchanged in the process. In our approach, the constraints are included in the objective function itself, thus making any restrictions on the placement of points an integral part of the algorithm. The proposed methodology is illustrated in several examples and its connection to model-based clustering is discussed.

CC135 Room Aula H MULTIVARIATE DATA ANALYSIS I — Chair: Friedrich Leisch

C0546: **A combined permutation test for comparing marginal probabilities of multivariate binary variables**

Presenter: **Stefano Bonnini**, University of Ferrara, Italy

Co-authors: Michela Borghesi

A multivariate extension of the two sample test on proportions is considered. We present a permutation solution based on the combination of the partial tests on the marginal probabilities of success. In particular, the testing problem concerns the comparison of two populations with respect to categorical data. The response is a multivariate binary variable. For such a problem, a nonparametric solution based on a statistic similar to that of the T-square Hotelling test was proposed in the literature for two-sided alternatives but there are no solutions for directional alternatives. Through a Monte Carlo simulation study, we prove the good performance of the test in terms of power. The results of the application of the proposed method to a case study concerning the propensity of companies toward a Circular Economy (CE) are also shown and discussed.

C0395: **Testing exchangeability of multivariate distributions**

Presenter: **Jan Kalina**, The Czech Academy of Sciences, Institute of Information Theory and Automation, Czech Republic

Although there have been a number of available tests of bivariate exchangeability, i.e. bivariate symmetry for bivariate distributions, the literature is void of tests on whether a multivariate distribution with more than two dimensions is exchangeable or not. Multivariate permutation tests of exchangeability of multivariate distributions are proposed, which are based on the nonparametric combination methodology, i.e. on combining nonparametric bivariate exchangeability tests. Numerical experiments on real as well as simulated multivariate data with more than two dimensions are presented here. The multivariate permutation test turns out to be typically more powerful than a bivariate exchangeability test performed only over a single pair of variables, and also more suitable compared to tests exploiting the approaches of Benjamini-Yekutieli or Bonferroni.

C0568: **On edge significance in large phylogenetic (spanning) trees**

Presenter: **Alexandre Francisco**, INESC-ID and IST, Universidade de Lisboa, Portugal

Co-authors: Pedro Monteiro, Guilherme Ribeiro, Andreia Sofia Teixeira

Tree-like structures are common in phylogenetic studies and, in intra-species studies with thousands of strains, minimum spanning trees (MST) and their generalizations are often computed. The underlying model assumes that edge weights (or just a total order) represent evolutionary distance. Given that MST are not unique in general, and some edges can be equally selected, it is important to compute their probability of being in an MST, i.e., their spanning edge betweenness (SEB). SEB can be computed exactly in $O(mn^{1.5})$ time (for n strains/vertices and m edges) through an extension of the well-known Kirchhoff's matrix tree theorem. The running time is however prohibitive for large datasets. But it turns out that SEB is equal to the effective resistance in electrical resistive networks and, for unweighted graphs, it can be provably approximated in $O\left(m\log^2(n)\log(1/\varepsilon)\right)$, by relying on provably fast linear system solvers for symmetric diagonally dominant (SDD) matrices. This result can be extended to weighted graphs when weights are probabilities and the MST weight is defined as the weight product, but its extension to the general case is not trivial. We propose then to address this problem by edge layering and combining partial results (as we did in the exact case), but controlling the approximation error through sampling and using the jackknife estimate of derived values.

C0569: **Towards the optimization of large-scale phylogenetic trees**

Presenter: **Catia Vaz**, INESC-ID / ISEL, Portugal
Co-authors: Alexandre Francisco

Several distance-based phylogenetic inference algorithms, widely used in the surveillance of infectious diseases, outbreaks investigation and studies of the natural history of infections, follow a hierarchical clustering approach to compute phylogenetic trees. Such algorithms differ in the similarity distance and in the optimization criteria used. Inferred trees might also not necessarily represent the best tree for the underlying evolution model. For instance, in the case of combinatorial optimization algorithms, such as goeBURST, that provide an optimal tree under a given criterion, we might not necessarily obtain the most representative phylogeny because distance does not always correlate with divergence time. And although we can further optimize trees using methods based on Subtree Pruning and Regrafting, Nearest Neighbor Interchange, or Tree Bisection and Reconnection, these methods are often expensive to compute, namely for large studies. We present then an extension of goeBURST that relies on efficient local optimizations to improve the inferred phylogeny, and which is applied to selected edges based on the maximum likelihood of two alternative evolutionary models. Underlying principles will be presented as well as results for both precision and sensitivity of the algorithm for reconstructing phylogenetic trees over simulated data.

Friday 26.08.2022 09:00 - 10:30 Parallel Session M – COMPSTAT2022

CV190 Room Aula H COMPUTATIONAL AND FINANCIAL ECONOMETRICS III Chair: Alessandra Amendola

C0648: The effect of climate policies on carbon emissions reduction
Presenter: **Jean-Baptiste Hasse**, Universite Catholique de Louvain, Belgium
Co-authors: Bertrand Candelon

The causal effects of climate policies on carbon emissions reductions are evaluated. Using Sweden as a case study, we compare the effects of the domestic carbon tax and the Kyoto protocol over the period 1965–2018. Specifically, we use the Granger-causality tests in the time and frequency domains to evaluate the impact of these climate policies. The empirical results indicate a significant causal effect in the long run of the carbon tax policy on carbon intensity dynamics. Finally, our framework offers policymakers a useful toolbox for evaluating the effect of public policy.

C0241: Multi-factor asset pricing model on functional principal component analysis
Presenter: **Bo Li**, Beijing International Studies University, China
Co-authors: Zhenya Liu, Shixuan Wang, Yifan Zhang

A functional principal component analysis (fPCA) procedure is proposed by incorporating non-linearity into estimating risk factor returns. It first converts cross-sectional returns sorted on univariate characteristics into functions to capture risk-induced non-linear changes in returns and then obtains the main statistical factors through fPCA. Using 205 characteristic-sorted returns, we verify that the first one is the market factor and propose using the eigenfunction of the second one as a weight vector to obtain fPCA factors. A further empirical study of 138 anomalies and 600 characteristic-sorted portfolios reveals that the fPCA factors improve the whole samples and the out-of-sample R2 by 10%-20%. The mean-variance efficiency portfolios of the fPCA factors achieve Sharpe ratios of more than 3.70, three times higher than the conventional factors.

C0266: Choosing between persistent and stationary volatility
Presenter: **Ilias Chronopoulos**, University of Essex, United Kingdom
Co-authors: Liudas Giraitis, George Kapetanios

A multiplicative volatility model is suggested where volatility is decomposed into a stationary and a non-stationary persistent part. We provide a testing procedure to determine which type of volatility is prevalent in the data. The persistent part of volatility is associated with a nonstationary persistent process satisfying some smoothness and moment conditions. The stationary part is related to stationary conditional heteroskedasticity. We outline theory and conditions that allow the extraction of the persistent part from the data and enable standard conditional heteroskedasticity tests to detect stationary volatility after persistent volatility is taken into account. Monte Carlo results support the testing strategy in small samples. The empirical application of the theory supports the persistent volatility paradigm, suggesting that stationary conditional heteroskedasticity is considerably less pronounced than previously thought.

C0612: Capital flows in integrated capital markets: The MILA case
Presenter: **Juan Vega Baquero**, Universitat de Barcelona, Spain
Co-authors: Miguel Santolino

The Feldstein-Horioka study on investment flows through the correlation of domestic saving and investment concluded that liberalization of capital markets does not necessarily lead to a movement of capital looking for a better allocation of resources, as classical theory would suggest. Ever since, literature has been prolific regarding this "puzzle", with arguments for and against this conclusion. The aim is to analyze the issue from a different perspective. In recent years, the stock markets of Chile, Colombia, Mexico and Peru joined the Latin American Integrated Market (MILA) through an agreement that allows investors in any of the participating markets to invest in the others as if they were investing locally. Compositional methods are used to assess the hypothesis of a potential flow of capital between markets generated by the creation of the joint market. Vector autoregressive models are estimated and tested for structural breaks in both, the parameters and the variance of the errors. As a result, it was not possible to find a change in the composition of the investment in the four markets produced by the creation of the joint market. Furthermore, compositional models showed to be more informative than the traditional ones in terms of the significance of the parameters and parsimony.

CV227 Room Aula I REGRESSION MODELS Chair: Peter Grunwald

C0633: Generalized score matching for regression
Presenter: **Jiazhen Xu**, Australian National University, Australia
Co-authors: Tao Zou, Andrew Wood, Janice Scealy

Many probabilistic models are developed with an intractable normalizing constant and have been extended to contain covariates. Since the evaluation of the exact full likelihood is difficult or even impossible for these models, score matching was proposed to avoid explicit computation of the normalizing constant. Score matching has been so far limited to the models in which the observations are independent and identically distributed (IID). However, the IID assumption does not hold in the traditional fixed design setting for regression-type models. To deal with the estimation of these covariate-dependent models, a novel score matching approach is presented for independent but not necessarily identically distributed(INID) data under a general framework for both continuous and discrete responses. In particular, we introduce a novel generalized score matching method for count response regression. We prove that our proposed score matching estimators are consistent and asymptotically normal under mild regularity conditions. The theoretical results are supported by numerical studies. Additionally, our simulation results indicate that, compared to the approximate maximum likelihood estimation, the generalized score matching produces estimates with smaller biases in an application to high school attendance data.

C0622: An approximation of the corrected naive estimator for a Poisson regression model with a measurement error
Presenter: **Kentarou Wada**, Tokyo University of Science, Japan
Co-authors: Takeshi Kurosawa

The corrected naive estimator is proposed as a consistent estimator for a Poisson regression model with a measurement error. The corrected naive estimator is given by the solution of a system of equations such as the moment method. The corrected naive estimator requires a tedious calculation to obtain the explicit form. Moreover, the corrected naive estimator does not always have an explicit expression under the condition that the explanatory variable and measurement error are general distributions. In this situation, we can compute the corrected naive estimator numerically even if the corrected naive estimator does not have an explicit solution. It takes some computational costs to obtain the numerical solution of the corrected naive estimator. We propose the approximate corrected naive estimator. The approximate corrected naive estimator has a simple closed form, which does not require a troublesome calculation. The approximate corrected naive estimator can be applied for the condition that the corrected naive estimator does not have an explicit expression. Furthermore, the bias of the approximate corrected naive estimator has a close value to that of the corrected naive estimator.

C0532: A weighted quantile sum regression with penalized weights and two indices
Presenter: **Stefano Renzetti**, Universita degli Studi di Brescia, Italy
Co-authors: Chris Gennings, Stefano Calza

An extension of Weighted Quantile Sum (WQS) regression is proposed which estimates the double effect of a mixture of chemicals on a health outcome in the same model through the inclusion of two indices, one in the positive and one in the negative direction, with the introduction of a penalization term. To evaluate the performance of this new model in terms of the estimation of the regression parameters and the weights we performed both a simulation study and a real case study where we assessed the effects of nutrients on obesity among adults. The results showed good performance of the method in estimating both the regression parameter and the weights associated with the single elements when the penalized term was set equal to the magnitude of the AIC of the unpenalized WQS regression. The two indices further helped to give a better estimate of the parameters (Positive direction Median Error (PME): 0.017; Negative direction Median Error (NME): -0.023) compared to the standard WQS (PME: -0.141; NME: 0.078). In the case study, WQS with two indices was able to find a significant effect of nutrients on obesity in both directions identifying caffeine and magnesium as the main actors in the positive and negative association respectively. We introduce an extension of the WQS regression that showed how to improve the accuracy of the parameter estimates when considering a mixture of elements that can have both a protective and a harmful effect on the outcome

C0407: **Divide and conquer approaches for nonparametric regression and variable selection**
Presenter: **Sapuni Chandrasena**, University of Toledo, United States
Co-authors: Rong Liu

The rapid emergence of massive data with increasing size requests new statistical methods, especially in the fields of nonparametric regression, which is flexible but usually computationally intensive. To overcome the limitations of computing and storage, various distributed frameworks for statistical estimation and inference have been proposed. We study the statistical efficiency and asymptotic properties of the spline estimation for generalized additive models using the divide-and-conquer (DAC) approach. We also provide a variable selection method based on the majority voting procedure. The simulation study strongly supports the asymptotic theory and shows that the DAC approach is much more computational expedient without losing much accuracy.

CO051 **Room Aula G** **IASC-ARS SESSION: COMPUTATIONS FOR CATEGORICAL DATA (VIRTUAL)** **Chair: Yuichi Mori**

C0292: **The Cressie-Read divergence statistic and correspondence analysis; a unifying approach with possible extensions**
Presenter: **Rosaria Lombardo**, University of Campania, Italy
Co-authors: Eric Beh

In the correspondence analysis literature, the foundations of visually and numerically summarising the association between two categorical variables rest with Pearson's chi-squared statistic. Not only is this statistic extremely popular and versatile, but it also yields some very useful visual and numerical properties. More recently, ties have been established that show the role that the Freeman-Tukey statistic plays in correspondence analysis and confirmed the advantages of the Hellinger distance that have long been advocated. Both Pearson's and the Freeman-Tukey statistics are special cases of the Cressie-Read divergence statistic, as are the Cressie-Read statistic, the likelihood ratio statistic and their modified versions. Therefore, correspondence analysis will be explored where the association, and the resulting low-dimensional correspondence plot, have at its foundation this divergence statistic. By doing so, the properties of correspondence analysis are described for any special case of the Cressie-Read divergences statistic which includes the Hellinger distance decomposition (HDD) method and log-ratio analysis (LRA). Some extensions to this method will also be discussed including its role in multiple and multi-way correspondence analysis.

C0369: **A multiple correspondence analysis for aggregated symbolic data**
Presenter: **Junji Nakano**, Chuo University, Japan
Co-authors: Nobuo Shimizu, Yoshikazu Yamamoto

When we have a huge amount of data, we sometimes are interested in comparing meaningful groups of data, not individual observations. Aggregated symbolic data (ASD) expresses a group of observations that have continuous and categorical variables by using up to second moments of variables. ASD for a group of data is equivalent to the set of means, variances, and correlations for continuous variables, Burt matrix for categorical variables, and means of a continuous variable against one value of a categorical variable. As ASD with many categorical variables is still complicated, it is preferable to have simple measures of location and dispersion for a categorical variable, and a measure of the correlation between two categorical and/or continuous variables. We propose such measures by extending multiple correspondence analysis to ASD. They are compared with other measures, for example, correlation measures based on the polychoric correlation coefficient.

C0489: **A general framework for implementing distance measures for categorical variables**
Presenter: **Michel van de Velden**, Erasmus University Rotterdam, Netherlands
Co-authors: Alfonso Iodice D Enza, Angelos Markos, Carlo Cavicchia

In many statistical methods, distance plays an important role. For instance, data visualization, classification and clustering methods require quantification of distances among objects. How to define such distance depends on the nature of the data and/or problem at hand. For the distance between numerical variables, in particular in multivariate contexts, there exist many definitions that depend on the actual observed differences between values. It is worth underlining that often it is necessary to rescale the variables before computing the distances. Many distance functions exist for numerical variables. For categorical data, defining a distance is even more complex as the nature of such data prohibits straightforward arithmetic operations. Specific measures, therefore, need to be introduced that can be used to describe or study the structure and/or relationships in the categorical data. We introduce a general framework that allows an efficient and transparent implementation of the distance between categorical variables. We show that several existing distances (for example distance measures that incorporate association among variables) can be incorporated into the framework. Moreover, our framework quite naturally leads to the introduction of new distance formulations as well.

C0544: **cGAPdb: A matrix visualization database for categorical data sets**
Presenter: **Chun-houh Chen**, Academia Sinica, Taiwan
Co-authors: Shao-An Chen, Chiun-How Kao, Sheau-Hue Shieh, Han-Ming Wu

cGAPdb is a graphical database for categorical data sets for public use. The major type of visualization provided in this database is matrix visualization with the cGAP (Categorical Generalized Association Plots) environment. Most of the categorical data sets from the UCI Machine Learning Repository are included in this graphical database. All elements of a cGAP display such as (homals analysis, data matrix, proximity matrix for variables and samples, seriation method, etc.) are provided for each data set for users to browse and download. Additional categorical data sets other than those from the UCI Repository have also been collected in cGAPdb. A cGAP working place is available in cGAPdb for users to upload their own data sets for creating cGAP matrix visualization displays.

CO037 **Room Aula C** **CLUSTERING METHODS AND COPULA FUNCTION** **Chair: F Marta L Di Lascio**

C0508: **Copula-based clustering of dependent variables with application to flood risks**
Presenter: **Roberta Pappada**, University of Trieste, Italy
Co-authors: Fabrizio Durante, Sebastian Fuchs

In recent years, copula-based measures of association have been exploited to develop clustering methods that can take into account the dependence structure characterizing the underlying data generating process, e.g., when the data objects to cluster are time series. Motivated by the interest in clustering flood data, which are characterized by a number of physical variables (such as flood peak and volume) and collected at specific geographical sites, some dissimilarity measures are proposed to cluster continuous random variables. Such measures are rank-based, hence depend on the copula of the involved random variables and assign the smallest value to two subsets of random variables that are pairwise comonotonic. Two different notions of multivariate comonotonicity for pairs of random vectors are investigated, which correspond to the strongest version of comonotonicity and a weaker notion called π-comonotonicity. The proposed dissimilarities are embedded into a hierarchical clustering procedure, with the final aim to detect clusters that account for the comovements of random variables. An application to the analysis of flood risks concerning data collected in the Po river basin is presented, along with the results from different simulated scenarios.

C0392: **Copula-based non-metric unfolding**

Presenter: **Marta Nai Ruscone**, Universita degli Studi di Genova, Italy

Co-authors: Antonio Dambrosio, Daniel Fernandez

A multidimensional unfolding technique that is not prone to degenerate solutions and is based on multidimensional scaling of a complete data matrix is proposed. We adopt the strategy of augmenting the data matrix, trying to build a complete dissimilarity matrix, by using Copulas-based association measures among rankings (the individuals), and between rankings and objects (namely, a rank-order representation of the objects through tied rankings). The proposed technique leads to an acceptable recovery of given preference structures. Applications on real datasets show that our procedure returns non-degenerate unfolding solutions.

C0218: **Mixtures with a prior on the number of components and the telescoping sampler**

Presenter: **Gertraud Malsiner-Walli**, WU Vienna University of Economics and Business, Austria

Co-authors: Sylvia Fruhwirth-Schnatter, Bettina Gruen

Within a Bayesian framework, a comprehensive investigation of the model class of mixtures of finite mixtures (MFMs) where a prior on the number of components is specified is performed. This model class has applications in model-based clustering as well as for semi-parametric density approximation, but requires suitable prior specifications and inference methods to exploit its full potential. We contribute to the Bayesian analysis of MFMs by (1) considering static and dynamic MFMs where the Dirichlet parameter of the component weights either is fixed or depends on the number of components, (2) proposing a flexible prior distribution class for the number of components, (3) characterizing the implicit prior on the number of clusters as well as partitions by deriving computationally feasible formulas, (4) linking MFMs to Bayesian non-parametric mixtures, and (5) finally proposing a novel sampling scheme for MFMs called the telescoping sampler which allows Bayesian inference for mixtures with arbitrary component distributions. The telescoping sampler explicitly samples the number of components, but otherwise requires only the usual MCMC steps for estimating a finite mixture model. The ease of its application using different component distributions is demonstrated on real data sets.

C0425: **Clustering Italian regions on the basis of bivariate income and consumption distributions**

Presenter: **Francesca Condino**, University of Calabria, Italy

Co-authors: Antonio Irpino, Rosanna Verde

In an economic framework, modeling income and consumption characteristics simultaneously can be of considerable relevance. Moreover, it could be of interest to identify homogeneous regions in a country in terms of economic behaviour. With this aim, we propose to jointly model income and consumption data through the copula approach and use the obtained bivariate density functions as descriptors of regions for clustering analysis purposes. In particular, considering data from the Survey on Households Income and Wealth (SHIW) by the Bank of Italy, the bivariate distributions at the regional level are obtained. The Jensen-Shannon divergence can be usefully employed to measure the discrepancies across density functions, as it allows us to take into account marginal and copula effects. The Italian regions are then partitioned in clusters by using a dynamic clustering algorithm, a non-hierarchical iterative algorithm, based on the optimization of an adequacy criterion that measures the fit between clusters and their prototypes. It can be shown that the divergence of all considered objects can be decomposed into two quantities, one relating to the heterogeneity present in the clusters and the other reflecting the discrepancy across clusters, according to Huygens' theorem.

CO148 **Room Aula D** EARLY CAREER ADVICE FOR STATISTICIANS IN THE COMPUTATIONAL SCIENCES **Chair: Thomas Yee**

C0295: **Different flavors of publishing computational work**

Presenter: **Ursula Laa**, BOKU University, Austria

The publication of work on computational methods comes in different flavors: from publishing software through a repository such as CRAN (with accompanying documentation), all the way to describing new concepts and approaches theoretically in a journal article. However, most of the time the ideal solution is somewhere in the middle: we both share the software and its documentation, and also describe the details in an associated research paper. We will present a broad overview of different types of journals relevant to computational statistics, and the accompanying expectations in terms of presentation and availability of software. This will be illustrated with examples from my own experience.

C0298: **Software development and statistical research: Some reflections**

Presenter: **Thomas Yee**, University of Auckland, New Zealand

Statisticians developing new methodology are obliged to provide software implementing the work as it facilitates its use and promotes reproducible research. However, writing good quality software takes much time, and this could be spent writing more papers. With fewer journal publications in general, academics pursuing this line of output are disadvantaged from those traditionally involved in publishing only. Some thoughts on this topic are shared. It is aimed more toward early-career researchers, however people of all ages should find something to identify with.

C0315: **The role of communities of practice for career development in computational statistics**

Presenter: **Laura Vana**, TU Wien, Austria

The aim is to reflect on how modern communities of practice, with a focus on meetup groups such as useR, R Ladies, PyLadies, can be leveraged by early career statisticians in the computational sciences to enhance their expertise, network and gain visibility. Finally, we will provide an overview of such modern communities in the area of computational statistics and data science as well as provide some guidelines on how to build and maintain impactful, safe and inclusive communities.

C0409: **What is the best programming language for computational sciences: No need to choose, be a polyglot**

Presenter: **Michele La Rocca**, University of Salerno, Italy

Early in their careers, a common question for students and data scientists is which programming language is best to learn. The question is somewhat misleading: every programming language has its strengths and weaknesses. Often, R and Python are compared with conclusions that, in some cases, point towards Python in others towards R. However, the correct answer to the question is not R *or* Python, but R *and* Python. Besides R and Python, Julia is receiving more and more attention from the data science community, again with significant strengths and some weaknesses. Especially at the beginning of their careers, computational scientists should be multilingual and learn complementary programming languages to cover future needs in different fields of application and career perspectives. The knowledge of any programming language is exposed to some degree of obsolescence. At the beginning of a career, the focus should not be on coding but rather on programming, especially on programming paradigms (OOP, functional programming, etc.) that have a higher degree of resilience.

CO035 **Room Aula F** RECENT DEVELOPMENTS OF VARIATIONAL APPROXIMATIONS **Chair: Mauro Bernardi**

C0464: **Variational Bayes for dynamic sparsity in time varying parameter regression with many predictors**
Presenter: **Nicolas Bianco**, University of Padova, Italy
Co-authors: Mauro Bernardi

Time-varying parameter models are powerful statistical tools for the analysis of dynamical systems. However, in high-dimensional problems the risk of over-parametrization is high, thus dynamic sparsity is desired. The latter is defined in two directions: vertical, where we look at the parameter vector at a fixed time, and horizontal, where we focus on a given variable and observe its behaviour across the timeline. We propose an extension of the Bernoulli-Gaussian model for variable selection to deal with time-varying sparsity by assuming a time dependence in the inclusion probabilities. We tackle the inference within a variational Bayes framework and we provide a global flexible approximation of the latent states exploiting a non-stationary Gaussian Markov random field representation. The properties of Bernoulli-Gaussian model together with the computational efficiency of variational methods enable a fast estimation and signal extraction also in regressions with many predictors.

C0481: **Bayesian non-conjugate regression via variational belief updating**
Presenter: **Cristian Castiglione**, University of Padova, Italy
Co-authors: Mauro Bernardi

A new variational algorithm is presented in order to provide a flexible tool for approximating the general posterior distribution of Bayesian models that combine subjective prior beliefs with an empirical risk function. Particular attention is delivered to regression and classification models linking data and parameters through a continuous convex loss function and a linear predictor. Many remarkable examples belonging to this class are of particular interest for statistical applications, such as generalized linear models, support vector machines, quantile and expectile regression. The proposed iterative procedure lies in the family of semiparametric variational Bayes and enjoys closed-form updating formulas along with efficient integration of the evidence lower bound. Neither conjugacy nor elaborate data augmentation strategies are required. Structured prior distributions, e.g., cross-random effects, spatial or temporal processes, inducing shrinkage and sparsity priors, can be easily accommodated into such a framework without additional effort since the modularity of mean field variational Bayes is preserved. The properties of the algorithm are then assessed through a simulation study and a real data application, where the proposed method is compared with Markov chain Monte Carlo and conjugate mean field variational Bayes in terms of posterior approximation accuracy, prediction error and computational runtime.

C0513: **Streamlined variational inference for sparse linear mixed model selection**
Presenter: **Luca Maestrini**, The Australian National University, Australia
Co-authors: Emanuele Degani, Dorota Toczydlowska, Matt P Wand

Variational approximations facilitate fast approximate Bayesian inference for the parameters of a variety of statistical models, including linear mixed models. However, for models with a high number of fixed or random effects, simple application of standard variational inference principles does not lead to fast approximate inference algorithms, due to the size of model design matrices and inefficient treatment of sparse matrix problems arising from the required approximating density parameters updates. We illustrate how recently developed streamlined variational inference procedures can be generalized to make fast and accurate inference for the parameters of linear mixed models with nested random effects and global-local priors for Bayesian fixed effects selection. Our variational inference algorithms achieve convergence to the same optima of their standard implementations, although with significantly lower computational effort, memory usage and time, especially for large numbers of random effects.

C0629: **Fast Bayesian model selection algorithms for linear regression models**
Presenter: **Manuela Cattelan**, University of Padova, Italy
Co-authors: Mauro Bernardi, Claudio Busatto

The issue of model selection for high-dimensional linear regression has been primarily addressed by assuming hierarchical mixtures as prior distributions. A spike component with Dirac probability mass at zero is introduced to exclude irrelevant covariates, thereby leading to Bayesian selection procedures that rely on the computation of the marginal posterior distribution for alternative model configurations. The exploration of the space of competing models is usually performed by means of computationally intensive simulation-based techniques. We address the issue of fast updating the variance-covariance matrix of the posterior distribution and the marginal posterior density itself, after a modification of the current design matrix. First, leveraging a thin QR factorization, novel algorithms to update the posterior variance-covariance matrix are proposed which avoid storage and update the Q matrix thus allowing noticeable savings. Then, the issue of evaluating the marginal posterior is considered, as it represents the bottleneck of any Bayesian model selection procedure. It is shown that the computation of the marginal posterior relies on the inverse of the R matrix, hence we develop a new methodology to update both this inverse and the related marginal posterior after the modification of the current design matrix. These methods do not need computationally intensive inversions of large dimensional matrices when performing marginal posterior evaluations.

CC231 Room Aula B TIME SERIES AND FINANCIAL ECONOMETRICS **Chair: Massimiliano Caporin**

C0692: **A convolutional approach to forecast reconciliation**
Presenter: **Andrea Marcocchia**, Sapienza University of Rome, Italy
Co-authors: Serena Arima, Pierpaolo Brutti

The goal of forecast reconciliation in a hierarchy is that observed responses/demands at each level will always add up to the observed responses/demands at higher levels. In the literature, there are numerous approaches that try to make predictions coherent: the challenge is to exploit the predictive power of the most aggregated data to benefit from it on the most granular data. The idea is to exploit the convolutions of Neural Networks to aggregate information at different levels of a hierarchy to obtain consistent and accurate predictions. In particular, we propose to generalize the convolutional approach to work in cases where more than one hierarchy is available, in order to make all the hierarchies simultaneously coherent. Another approach, in which convolutions are exploited on a graph data structure instead of a matrix, has been tested and it is currently under development. The advantage of this approach is that it is not needed to use a rigid data structure such as a matrix, but it is possible to exploit the flexibility of graphs. Several benchmark datasets are being tested, both simulated and real.

C0694: **Fundamental and speculative components of the cryptocurrency pricing dynamics**
Presenter: **Jiri Kukacka**, Czech Academy of Sciences, Institute of Information Theory and Automation, Czech Republic
Co-authors: Ladislav Kristoufek

The driving forces behind crypto assets price dynamics are often perceived as being dominated by speculative factors and inherent bubble-bust episodes. The fundamental components are believed to have a weak, if any, role in price formation and emergent dynamics. This research studies five crypto assets with different backgrounds, including Bitcoin, Ethereum, Litecoin, XRP, and Dogecoin between 2016 and 2022. It utilizes the cusp catastrophe model to connect the fundamental and speculative drivers with possible price bifurcation characteristics of events of a market collapse. The findings show that all studied assets except Dogecoin demonstrate their price and return dynamics emerge from complex interactions among both fundamental and speculative components, including episodes of pricing bifurcations. Bitcoin shows the strongest fundamentals, with the on-chain activity driving the fundamental part of the dynamics. Investor attention mainly drives the speculative component for all studied assets. Thus, the fundamental factors should not be left out when constructing crypto assets pricing models.

C0683: **Instability in SETAR models**
Presenter: **Pu Chen**, Melbourne Institute of Technology, Australia

The focus is on the instability of a self-exciting regime-switching autoregressive model, specifically regime-switching models that are locally stable in each of their regimes. It turns out that the local stability of each regime is insufficient to ensure the overall stability of the model. The mechanism leading to the instability is described, and a sufficient condition for the instability is provided. The developed condition is then extended to STAR model and Markov switching models.

C0700: **Realized stochastic volatility models with skew-t distributions**

Presenter: **Makoto Takahashi**, Hosei University, Japan

Co-authors: Yasuhiro Omori, Toshiaki Watanabe, Yuta Yamauchi

Predicting volatility and quantiles of financial returns is essential to measure the financial tail risk such as value-at-risk and expected shortfall. There are two important aspects of volatility and quantile forecasts: the distribution of financial returns and the estimation of the volatility. Building on the traditional stochastic volatility model, the realized stochastic volatility model incorporates realized volatility as the precise estimator of the volatility. Using three types of skew-t distributions, the model is extended to capture the well-known characteristics of the return distribution, namely skewness and heavy tails. In addition to the normal and Student's t distributions, included as the special cases of the skew-t distributions, two of them contain the skew-normal, and hence allow more flexible modeling of the return distribution. The Bayesian estimation scheme via a Markov chain Monte Carlo method is developed and applied to major stock indices. The empirical study using the US and Japanese stock indices data suggests that incorporating both skewness and heavy tail to daily returns is important for volatility and quantile forecasts.

CC210 Room Aula Q GRAPHICAL MODELS AND NETWORKS **Chair: Mohammad Arashi**

C0529: **Estimation and inference for covariate-adjusted Gaussian graphical models via an unbalanced distributed setting**

Presenter: **Ensiyeh Nezakati Rezazadeh**, Universita catholique de Louvain (Belgium), Belgium

Co-authors: Eugen Pircalabelu

Precision matrix estimation plays an important role in statistical and machine learning framework, especially in the framework of Gaussian graphical modeling. Most current methods for precision matrix estimation assume that the random vector has zero or constant mean. However, in many real applications, like genomic data analysis, it is often important to adjust for the covariate effects on the mean of the random vector to obtain more precise estimates. On the other hand, in the estimation framework, many modern datasets are characterized by both large dimensions and sample size, such that they cannot be stored in one single machine. In this vein, new algorithms have been developed for splitting a dataset on different local machines with different capacities such that the estimation will be solvable in each machine. New unbalanced distributed estimations are provided for both the covariate mean and precision matrices in adjusted-covariate Gaussian graphical models. These estimators aggregate all local estimators into the final ones by maximizing the pseudo loglikelihood function which comes from the asymptotic distribution of debiased estimators in the subsamples. Asymptotic behavior and statistical guarantees of these estimators are provided when the number of parameters, covariates and machines all grow with the sample size. A simulation study and a real data example are used to assess the performance of these estimators.

C0627: **A catalogue of graph-based multivariate conditional autoregressive model**

Presenter: **Anna Freni Sterrantino**, The Alan Turing, United Kingdom

Co-authors: denis rustand, Haavard Rue

An intuitive approach is presented to define a Multivariate conditional autoregressive model(MCAR) based on graphs and using Kronecker models. The MCAR precision is given as the Kronecker product of the inverse of a correlation matrix and the precision of an Intrinsic Conditional autoregressive model, representing the spatial structure. We frame the MCAR models into the application of multivariate disease mapping and to represent different correlations structures, we have created a catalogue of graphs to model up to four variables (diseases) and introduced the penalized complexity priors as hyper-priors for this parametrization. The penalized complexity priors penalize departures from a model with independent variables and pure overdispersion (base model) compared to a complex model with correlation among diseases and structured spatial variability. The resulted priors are weakly informative and shrink the correlation parameters toward zero. These models find their main application in epidemiology but can be easily extended to other fields of applications.

C0246: **Analysing populations of networks with mixtures of generalized linear mixed models**

Presenter: **Mirko Signorelli**, Leiden University, Mathematical Institute, Netherlands

Until recently, collecting data on populations of networks (PoN) was cumbersome and rare. However, the increasing availability of automatic monitoring devices and the growing scientific interest in networks make such data more widely available. From sociological experiments involving cognitive social structures to fMRI scans revealing large scale brain networks of groups of patients, there is growing awareness that we urgently need statistical methods to analyse and summarize PoN. We will show how mixtures of generalized linear mixed models can be employed to model PoN in a thrifty but interpretable way. This model-based clustering approach to PoN allows identifying subpopulations of networks that share certain topological properties (degree distribution, community structure, effect of covariates on the presence of an edge, etc.) of interest. We will discuss how the proposed model can be estimated by combining adaptive gaussian quadratures with the EM algorithm and assess its classification performance using simulated data. We will conclude by illustrating an example application of the proposed method to a PoN representing how employees perceive advice relationships within a small business, paying particular attention to model specification and the interpretation of the outputs of our model.

C0523: **Flexible parametrization of graph-theoretical features from individual-specific networks for prediction**

Presenter: **Mariella Gregorich**, Medical University of Vienna, Austria

Nowadays, many structurally similar predictors are available for each individual, which can be represented as individual-specific networks able to capture their dependence structure and can provide predictive biomarkers for outcome modelling. However, unsubstantiated, arbitrary decisions in individual-specific network inference, in particular when choosing a suitable threshold for network sparsification, still lead to a high variability of the extracted graph-theoretical features. We propose a flexible parameterization approach to include graph-theoretical features as explanatory variables in a prediction model. In particular, flexible functional weight functions of the threshold value determined by statistical goodness-of-fit criteria enable us to incorporate uncertainties of network inference in the model. We perform a simulation study to provide evidence for a proof-of-concept in individual-specific networks of a given size, density and with particular, well-defined network properties. We compare the predictive performance of our approach to a more conventional method of selecting a single sparsification threshold based on AIC. We highlight some challenges that need to be addressed before our approach is ready for routine applications and provide recommendations for our proposed approach in an applied data setting.

CC228 Room Aula E MULTIVARIATE DATA ANALYSIS II **Chair: Alina Matei**

C0671: **Transformation and covariance estimation for the non-linearly separable misclassification problem**

Presenter: **Mubarak AL-Shukeili**, Sultan Qaboos University, Oman

Co-authors: Ronald Wesonga

The search for a suitable binary classifier mainly depends on the location vector and covariance matrix. The linear discriminant, for example, works by constructing the most suitable linear hyperplane based on location and covariance among parameters. The data point X are transformed from

R^p to R such that the resultant data leads to a minimum misclassification rate. However, when the class data are significantly overlapped due to class homogeneity, linear classifiers perform poorly. We present a method that performs the transformation of classes to be linearly separable prior to classification. Moreover, our study also proposed an estimate of a covariance matrix for one class given the other class is known. The resultant performance of the proposed method is validated using simulation and real-life data. Findings show that our method yields more competitive results compared to the classical quadratic discriminant analysis

C0387: **Blind source separation for multivariate stationary space-time data**

Presenter: **Christoph Muehlmann**, Technical University of Vienna, Austria

Co-authors: Sandra De Iaco, Klaus Nordhausen

With advances in modern world technology, huge datasets that show dependencies in space as well as in time occur frequently in practice. As an example, several monitoring stations at different geographical locations track hourly concentration measurements of a number of air pollutants for several years. Such a dataset contains thousands of multivariate observations, thus, proper statistical analysis needs to account for dependencies in space and time between and among the different monitored variables. To simplify the consequent multivariate spatio-temporal statistical analysis it might be desirable to find linear transformations of the original data that result in easy interpretative, spatio-temporally uncorrelated processes that are also highly likely to have real physical meaning. Blind source separation (BSS) is a statistical methodology that is concerned with finding so-called latent processes that exactly meet the former requirements. BSS was already successfully used for sole temporal and sole spatial data, but, it was not yet introduced for the spatio-temporal case. BSS is reviewed and a generalization of BSS for second-order stationary multivariate spatio-temporal random fields (stBSS) is proposed. Two novel estimators (stAMUSE and stSOBI) which solve the formulated problem are also provided.

C0333: **Sparse principal loading analysis**

Presenter: **Jan Bauer**, University of Basel, Switzerland

Principal loading analysis is a dimension reduction method for cross-sectional data that selects a subset of the existing variables. Variables that have a small distorting effect on the covariance matrix are discarded. However, the method considers a hard threshold rule for the eigenvectors of the covariance matrix in order to determine if variables have a small distorting effect. We contribute an extension to sparse principal loading analysis, where we rather work with sparse loadings than with threshold rules for the eigenvectors. The sparse loadings are obtained using penalization methods and we compare calculation approaches regarding their explained variance which is needed to evaluate if a variable is selected or not. Further, we discuss sparse principal loading analysis in contrast to principal loading analysis and we contribute applications to real and simulated data as an illustration.

C0463: **Connecting compositional data to graph signal processing**

Presenter: **Christopher Rieser**, TU Wien, Austria

Traditional methods for the analysis of compositional data consider the log-ratios between all different pairs of variables with equal weight, typically in the form of aggregated contributions. This is not meaningful in contexts where it is known that a relationship only exists between very specific variables (e.g. for metabolomic pathways), while for other pairs a relationship does not exist. Modeling the absence or presence of relationships is done in graph theory, where the vertices represent the variables, and the connections refer to relations. We show how to link compositional data analysis with graph signal processing and extend the Aitchison geometry to a setting where only selected log-ratios can be considered. The presented framework retains the desirable properties of scale invariance. An example from bioinformatics is shown to demonstrate the usefulness of this approach.

Friday 26.08.2022 11:00 - 12:00 Parallel Session N – COMPSTAT2022

CV202 Room Aula G SURVIVAL ANALYSIS (VIRTUAL) Chair: Martina Mittlboeck

C0408: **Estimation in the Cox survival regression model with covariate measurement error and a changepoint**
Presenter: **Sarit Agami**, The Hebrew University of Jerusalem, Israel

The Cox regression model is a popular model for analyzing the relationship between a covariate vector and a survival endpoint. The standard Cox model assumes a constant covariate effect across the entire covariate domain. However, in many epidemiological and other applications, the covariate of main interest is subject to a threshold effect: a change in the slope at a certain point within the covariate domain. Often, the covariate of interest is subject to some degree of measurement error. We study measurement error correction in the case where the threshold is known. Several bias correction methods are examined: two versions of regression calibration (RC1 and RC2, the latter of which is new), two methods based on the induced relative risk under a rare event assumption (RR1 and RR2, the latter of which is new), a maximum pseudo-partial likelihood estimator (MPPLE), and simulation-extrapolation (SIMEX). We develop the theory, present simulations comparing the methods, and illustrate their use on data concerning the relationship between chronic air pollution exposure to particulate matter PM10 and fatal myocardial infarction (Nurses Health Study (NHS)), and on data concerning the effect of a subjects long-term underlying systolic blood pressure level on the risk of cardiovascular disease death (Framingham Heart Study (FHS)). The simulations indicate that the best methods are RR2 and MPPLE.

C0410: **Improvement of midpoint imputation for estimation of median survival time for interval-censored time-to-event data**
Presenter: **Yuki Nakagawa**, Chugai Pharmaceutical Co., Ltd., Japan
Co-authors: Takashi Sozu

Progression-free survival (PFS) is used to evaluate a treatment effect for patients with solid tumors in cancer clinical trials. The disease progression of the patients is typically determined by radiological testing at several scheduled tumor-assessment time points. This results in a discrepancy between the true progression time and the observed progression time. Considering the observed progression time as the true progression time, a biased PFS is obtained for some patients, and the estimated survival function from the Kaplan-Meier method is also biased. Although the midpoint imputation method replaces the interval-censored data with the midpoint data and it reduces the bias, it has an unrealistic assumption that several disease progressions occur at the same time point when several disease progressions are observed in the same tumor-assessment interval. We improved the midpoint imputation method, which replaces the interval-censored data with the equally spaced timepoint data based on the number of observed interval-censored data within the same tumour-assessment interval. We evaluated the bias, root mean square error of the median, and coverage probability of the 95% confidence interval of the proposed method. The proposed method provided higher performances compared with those of the midpoint imputation method.

C0656: **Simulating survival data for cure models in overall or net survival framework**
Presenter: **Juste Goungounga**, University of Burgundy, INSERM, UMR1231, France
Co-authors: Olayide Boussari, Valerie Jooste

Simulation studies are pivotal in survival analysis to evaluate the performances of new and existing models. When the cure assumption holds, cure models allow estimating either overall survival or net survival (survival that would be observed if the studied disease -say cancer- were the only possible cause of death) and its asymptotic value, the cure fraction. Net survival is usually estimated by splitting the observed mortality into two forces: one due to cancer (excess mortality) and one due to other causes (expected mortality). Hence, a mechanism to generate survival data for the cure model in a net survival framework must include two independent time variables: time until death by cancer and time until death by other causes. To reflect plausible real data, a specified cure model for data generation could also require flexible and complex functions. We present methods for using different distributions when generating survival times by varying time-to-cure and cure fractions. We illustrate a numerical integration approach that could be used when a closed-form of cumulative hazard due to cancer does not exist and root finding techniques as an alternative to simulate survival times from mixture and non-mixture cure models. A user-friendly R package is also provided.

CV186 Room Virtual Room R1 CLUSTERING AND CLASSIFICATION I (VIRTUAL) Chair: Tsung-I Lin

C0543: **Fuzzy cluster-scaled principal component analysis for high-dimension low-sample data**
Presenter: **Mika Sato-Ilic**, University of Tsukuba, Japan

A study of the fuzzy clustering-based Principal Component Analysis (fuzzy clustering-based PCA) is presented which is capable of treating high-dimension, low-sample size data (HDLSS data) with high performance compared with ordinary PCA. In general, HDLSS data has difficulty analyzing by using conventional data reduction methods such as an ordinary PCA due to the inconsistency of eigenvalues of the sample covariance matrix with respect to variables, although one of the purposes of this analysis is the reduction of the number of dimensions (variables). In addition, in fuzzy clustering, the status of the clustering result shows not only whether the object belongs to clusters but also how much the object belongs to clusters. This can consider the practical situation of data. Therefore, the fuzzy clustering-based PCA can tackle the problem of ordinary PCA for the HDLSS data by including the scale of the result of fuzzy clustering. In particular, an application of this fuzzy clustering-based PCA is shown for the discrimination of individual subjects observed by sensors worn on the body during several activities. The analysis of this data is useful for healthcare, considering the individuality of the history of activities.

C0547: **A hybrid cross entropy method for spatial clustering problems**
Presenter: **Nishanthi Raveendran**, Macquarie University, Australia
Co-authors: Georgy Sofronov

Spatial clustering is one of the important components of spatial data analysis. Spatial data are often heterogeneous, indicating that there may not be a unique simple statistical model describing the data. However, if we cluster the data into homogeneous clusters or domains, it will be easier to apply the appropriate statistical model for each domain. The problem of finding homogeneous domains is known as segmentation, partitioning or clustering. It is commonly used in many areas including disease surveillance, spatial epidemiology, population genetics, landscape ecology, crime analysis and many other fields. We focus on identifying homogeneous clusters and their boundaries in spatial data which is commonly used in epidemiological applications. To solve this clustering problem, we propose to combine the Cross-Entropy method, which is one of the evolutionary computing techniques that utilize a stochastic framework to solve estimation and a variety of optimization problems, with Voronoi tessellation to estimate the boundaries of such domains. Our results illustrate that the proposed algorithm is effective in identifying homogeneous clusters in spatial data.

C0555: **Confidence interval for recall and precision of multi-class classification**
Presenter: **Kanae Takahashi**, Hyogo College of Medicine, Japan
Co-authors: Kouji Yamamoto

In the medical field, binary classification problems are common, and accuracy, sensitivity, specificity, negative predictive value, and positive predictive value are often used as indicators of the performance of binary predictors. Also, in computer science, classifiers are usually evaluated with recall (sensitivity) and precision (positive predictive value). Recall and precision are only applicable to binary classification data. Two aggregate performance measures have been proposed for recall and precision in multi-class classification problems: macro-averaged recall and precision (maR, maP) and micro-averaged recall and precision (miR, miP). The maR is the arithmetic mean of recall for each class and maP is the arithmetic mean of precision for each class. On the other hand, miR and miP are the recall and precision computed from the sum of the decisions per sample. Most articles report point estimates of recall and precision for multi-class classification without considering the uncertainty of the estimates. Therefore, we propose methods for estimating maR, maP, miR, and miP with confidence intervals based on the large sample multivariate central limit theorem and the delta method.

CV199 **Room Aula C** MULTIVARIATE DATA ANALYSIS (VIRTUAL) **Chair: Sonja Greven**

C0256: **Two-stage target rotation with computational efficiency by asymmetric least squares criterion**
Presenter: **Naoto Yamashita**, Kansai University, Japan

In factor analysis, a factor loading matrix is often rotated to a simple target matrix for its simplicity. As such rotational procedures, Promax and Simplimax are commonly used. A well-known limitation of Simplimax rotation is the computational inefficiency in estimating the sparse target matrix, which yields a considerable number of local minima. The target rotation procedures approximate the non-zeros in the loading matrix to zeros or non-zeros in the target matrix, but the existing procedures equally treat the two types of approximation, while the former is of importance for simplifying the loading matrix. The research proposes a new rotation procedure that consists of the following two stages. The first stage estimates a sparse target matrix with lesser computational cost by a regularization technique. In the second stage, a loading matrix is rotated to the target, emphasising the approximation of non-zeros to zeros in the target by the asymmetric least squares criterion. The simulation study and real data examples showed that the proposed method simplifies loading matrices, and its performance is superior to the existing procedures.

C0565: **Sensitivity analysis of the choice of multiple imputation approach on categorical GPAbin biplots**
Presenter: **Johane Nienkemper-Swanepoel**, Stellenbosch University, South Africa
Co-authors: Niel Le Roux, Sugnet Lubbe

Multiple imputation is generally considered as the recommended approach for the handling of missing data. This approach entails the computation of multiple completed data sets which are then analyzed separately by means of standard complete data techniques. Estimates from the separate analyses are combined using suitable combining rules. A popular method for combining the estimates is the so-called Rubin's rules. In the context of exploratory analysis, GPAbin biplots enable the combined visualization of multivariate visualizations constructed from multiple imputed data sets. This visualization approach combines configurations by means of generalized orthogonal Procrustes analysis (GPA) and applying Rubin's rules (-bin) to the aligned configurations. The performance of the GPAbin biplots has been evaluated using multiple imputation with multiple correspondence analysis (MIMCA) in an extensive simulation study. The performance of GPAbin will be evaluated when applying various multiple imputation techniques available in R. The effect of the choice of the multiple imputation approach will be investigated and presented by means of simulated examples.

C0550: **Power transformation of reciprocal averaging**
Presenter: **Ting-Wu Wang**, University of Newcastle, Australia
Co-authors: Eric Beh

A common difficulty when analysing categorical variables of a contingency table is that cell frequencies are often prone to overdispersion as they are typically assumed to be Poisson random variables. Transformation of data has been a popular and successful technique in statistics to overcome the issues introduced due to the equivalence feature between the expected value and the variance. A power transformation of the frequencies of the contingency table is therefore a strategy that can be used to help overcome this issue in the application of correspondence analysis. The traditional approach to correspondence analysis and the method of reciprocal averaging have close links in the analysis of the association between categorical variables. The connections between correspondence analysis and the method of reciprocal averaging are illustrated in the context of power transformations. Applications are also used to demonstrate the algorithms between how the two methods acquire solutions that are a weighted average of one another. In the calculation of the scores and the association between them, two alternative ways of choosing an appropriate power parameter are considered. One approach calculates the power parameter that explains the optimal association as a percentage of the phi-squared statistic from the first dimension of the solution. The second method involves determining a power transformation, such that the data becomes non-dispersed.

CI099 **Room Aula F** DATA VISUALIZATION AND MODEL SELECTION **Chair: Christophe Croux**

C0673: **Data and model visualisation for statistical learning problems**
Presenter: **Catherine Hurley**, Maynooth University, Ireland

Visualization techniques assist in getting to know data prior to modelling, and post-modelling, in exploring and comparing fits, diagnosing lack of fit and understanding predictor effects. We give an overview of some techniques we have developed for visualization in the context of statistical learning, which help address the interpretability deficit. We describe improved methods for exploring predictor importance, predictor interaction and partial dependence model summaries. We use interactive visualisation to dig deeper into model fits by focusing on slices of predictor space, thus investigating local lack of fit, local predictor effects and higher-order interactions. Our techniques are model agnostic and are appropriate for any regression or classification problem. The methods presented are implemented in the R packages vivid and condvis2.

C0645: **Data inspection via challenging decision boundaries' rigidity**
Presenter: **Anthea Merida**, Ecole Normale Superieure Paris Saclay, France
Co-authors: Argyris Kalogeratos, Mathilde Mougeot

How smooth decision boundaries are needed in order to better fit a certain dataset? Answering this question can be useful when analyzing a dataset. It can provide insight into the dataset itself, and can also help reduce the scope of exploration of the subsequent model selection procedure for a task at hand. To answer this question, we propose the quantification of how much given 'rigid' decision boundaries (produced by an algorithm that naturally finds such solutions) should be relaxed to achieve a performance improvement. The procedure starts with the rigid decision boundaries of a seed Decision Tree (DT), that is used to initialize a Neural DT. The latter is a Neural Network that is built using a DT, and whose activation functions' smoothness can be controlled by a hyperparameter. The boundaries are challenged by relaxing them progressively, through smoothing the NDT's activation functions and further training. During the procedure, the NDTs performance and decision agreement to its seed DT are measured. These two measures, along with the value of the smoothness parameter, are shown to be helpful for the user in figuring out how expressive their model should be, before exploring it further via model selection. The validity of this approach is demonstrated with experiments on simulated and other benchmark datasets.

CO075 **Room Aula D** COMPUTATIONAL STATISTICS FOR APPLICATIONS **Chair: Marta Disegna**

C0285: **Estimating the susceptible component of a zone diameter distribution**
Presenter: **Bettina Gruen**, WU (Vienna University of Economics and Business), Austria
Co-authors: Helga Wagner, Thomas Petzoldt

Disk diffusion tests are employed in diagnostic laboratories to determine the susceptibility of bacteria to antibiotics. The bacteria, as well as the antibiotics, are administered to an agar plate and the zone diameter (ZD) of inhibition is measured. ZD data have restricted support and usually, only rounded values are observed. Previous work suggested using a composite model of a parametric distribution covering the range of observations from the susceptible component combined with a non-parametric distribution capturing the range of observations containing also resistant observations when modeling the minimum inhibitory concentration that separates the susceptible from the resistant bacterial sub-population. We investigate the use of this model for ZD data and consider in addition a two-component mixture model of a parametric distribution capturing the susceptible observations and a non-parametric distribution for the resistant observations. We present maximum likelihood estimation of both models, the composite as well as the mixture model, for arbitrary parametric distributions taking the restricted support and the rounding of the data into account and outline the computational tools required for implementing the estimation as well as additional inference methods for model selection, visualization and estimation of a data-driven epidemiological cut-off.

C0340: **A multivariate permutation test for the analysis of market research data**
Presenter: **Marta Disegna**, University of Padova, Italy
Co-authors: Riccardo Ceccato, Rosa Arboretti, Elena Barzizza, Nicolo Biasetton, Luca Pegoraro, Luigi Salmaso

The Nonparametric Combination (NPC) is a flexible methodology that can be adopted to deal with a wide range of complex scenarios. It allows us to propose quite powerful testing procedures to undertake problems involving the comparison of two populations when a multivariate outcome is observed. We propose a new NPC-based test to deal with a particular multivariate scenario in which two paired samples and multiple data types are available. This technique is then adopted for the analysis of market research data. A questionnaire was indeed submitted to multiple respondents, asking them to evaluate a product in terms of a certain set of KPIs (i.e. ordinal variables were gathered) after two different time frames. A number of experiments were also conducted and additional KPIs were measured (i.e. numeric variables were collected) after the same two-time frames. The NPC-based test was therefore adopted to compare the performances of the product across time.

C0527: **INDCLUS for spatial proximity data**
Presenter: **Laura Bocci**, Sapienza University of Rome, Italy
Co-authors: Pierpaolo Durso, Vincenzina Vitale

A suitable extension of the INDCLUS model is proposed for clustering spatial units in three-way proximity data taking into account the spatial nature of the units. Specifically, our concern is three-way two-mode data consisting of square symmetric matrices $\mathbf{S}\_h$ $(h = 1,\ldots,H)$ of pairwise proximities of a set of I spatial units coming from H domains. INDCLUS searches for a covering of the units, which is common to all the H domains, and a set of weights and an additive constant, which are different for each domain. The model is fitted by solving a least-squares optimization problem. In order to identify a covering of spatial units accounting for and taking advantage of the spatial nature of the units themselves, a penalty term based on a suitable spatial contiguity matrix of size I is added to the objective function. Furthermore, a tuning coefficient allows to balance the identification of both a common classification of units for all domains and approximately spatial homogeneous clusters. An Alternating Least-Squares algorithm is provided to solve the penalized problem. The proposed method has been applied to the subset of BES indicators included in the Economic and Financial Document (DEF), submitted annually to the Government and approved by Parliament.

CO174 Room Aula H GEOSTATISTICS **Chair: Pier Giovanni Bissiri**

C0326: **A model for space-time threshold exceedances**
Presenter: **Carlo Gaetan**, Ca' Foscari University of Venice, Italy
Co-authors: Paola Bortot

In the context of space-time data, the challenge is to develop models for threshold exceedances that account for both spatial and temporal dependence. We address this issue through a modelling approach that embeds spatial dependence within a time series formulation. The model allows for different forms of limiting dependence in the spatial and temporal domains as the threshold level increases. In particular, temporal asymptotic independence is assumed, as this is often supported by empirical evidence, especially in environmental applications, while both asymptotic dependence and asymptotic independence are considered for the spatial domain. Inference from the observed exceedances is carried out through a combination of pairwise likelihood and a censoring mechanism. For those model specifications for which direct maximization of the censored pairwise likelihood is unfeasible, we propose an indirect inference procedure. The approach is applied to a dataset of rainfall amounts.

C0357: **Asymptotic properties of pseudo-ML estimators based on covariance approximations**
Presenter: **Reinhard Furrer**, University of Zurich, Switzerland
Co-authors: Michael Hediger

Maximum likelihood (ML) estimators for covariance parameters are highly popular in inference for random fields. In the years, the dataset sizes have steadily increased such that ML approaches can become quite expensive in terms of computational resources. Several covariance approximation approaches have been proposed (e.g., tapering, direct covariance misspecification, low-rank approximation) and have various advantages and disadvantages. We present an approach based on covariance function approximations that are not necessarily positive definite functions. More specifically, for a zero-mean Gaussian random field with a parametric covariance function, we introduce a new notion of likelihood approximations (termed pseudo-likelihood functions), which complements the covariance tapering approach. Pseudo-likelihood functions are based on direct functional approximations of the presumed covariance function. We show that under accessible conditions on the presumed covariance function and covariance approximations, estimators based on pseudo-likelihood functions preserve consistency and asymptotic normality within an increasing-domain asymptotic framework.

C0675: **Positive definite functions on spheres: Some statistical and mathematical issues**
Presenter: **Pier Giovanni Bissiri**, -, Italy
Co-authors: Emilio Porcu, Felipe Tangle, Ruben Soza, Fernando Quintana

Positive definite functions are a key mathematical tool in geostatistics. If the study region is geographically extensive, the space which needs to be considered is the spherical surface equipped with the geodesic distance. After reviewing some general results about positive definite functions on spheres, the focus is on Bayesian nonparametric models for spatial covariance functions for global data. Then, it will review the main known results about strict positive definiteness on spheres in the isotropic case and will show a recent result in the axially symmetric case.

CO065 Room Aula I RESEARCH METRICS FOR INSTITUTIONAL PERFORMANCE EVALUATION (VIRTUAL) **Chair: Keisuke Honda**

C0460: **Development of visualizing system based on research networked data for open science age**
Presenter: **Hiroka Hamada**, The Institute of Statistical Mathematics, Japan
Co-authors: Keisuke Honda

To see how papers are influencing different fields, an index has been developed that defines the degree of heterogeneity based on only citation relations and scores the scattering of citations. In general, clustering of network structures, like stochastic block models, requires computing the entire data. Non-negative matrix factorization is applied to estimate the global citation structure of the bibliography from a small subset by sampling. This allows the system to be offered as a license-free product using various open science services. We will introduce the system is provided with a 3D plot which is based on kernel PCA to easy understand the multidimensional assignments of the scores of the diversity indexes. Furthermore, a use case for the administrator of an institution to analyze research resources by linking this system with funding information is discussed in our demo.

C0286: **A leading author model for the popularity effect on scientific collaboration**
Presenter: **Frederick Kin Hing Phoa**, Academia Sinica, Taiwan

Co-authors: Hohyun Jung, Mahsa Ashouri

The focus is on the popularity effect of the scientific collaboration process that popular authors have an advantage in making more publications. Standard network analysis has been used to analyze the scientific collaboration network. However, the standard network has limitations in explaining the scientific output by binary co-authorship relationships since papers have various numbers of authors. We propose a leading author model to understand the popularity effect mechanism while avoiding the use of the standard network structure. The estimation algorithm is presented to analyze the size of the popularity effect. Moreover, we can find influential authors through the estimated genius levels of authors by considering the popularity effect. We apply the proposed model to the real scientific collaboration data, and the results show positive popularity effects in all the collaborative systems. Furthermore, finding influential authors considering the genius level is discussed.

C0355: **Assessing the research strength of organizations focusing on intrapersonal diversity in applied research of AI**

Presenter: **Yuji Mizukami**, Nihon University, Japan

Co-authors: Junji Nakano

The focus is on "joint research between different research fields" and discussing how each country promotes research from an innovation perspective by classifying the styles of cross-disciplinary integration into several patterns. The methods for analysis provide a measurable framework for the concept of "intrapersonal diversity" in the innovation strategy of Schumpeterian competition, and provide an example of the application of the management theory "ambidextrous management" to induce innovation. The information on intrapersonal diversity can be gathered and evaluated as an organization's competitive strength. For the analysis, we used 19-year bibliographic data (2000-2018) from the top 20 countries in terms of the number of papers in AI technology. The data were processed using the co-authorship analysis method proposed by the authors and the newly presented cross-disciplinary collaboration display method. As a result, the styles of cross-disciplinary fusion are categorized into four patterns in AI.

CO127 **Room Aula Q** MATHEMATICAL AND STATISTICAL METHODS FOR ECONOMICS AND FINANCE **Chair: Luca Vincenzo Ballestra**

C0467: **Pricing options using a score-driven model with jumps**

Presenter: **Luca Vincenzo Ballestra**, Alma Mater Studiorum University of Bologna, Italy

Co-authors: Enzo DInnocenzo, Andrea Guizzardi

Score-driven models can provide significant improvements over GARCH models in fitting and forecasting asset prices. We present a score-driven model with jumps (SDJ) for option pricing. In particular, the conditional variance of the returns is specified by an autoregressive process driven by the score of the predictive density, whereas jumps follow a compound Poisson process. This allows us to consider the interaction between jumps and volatility, as the Poisson process and the dynamics of the variance turn out to be fully coupled. Furthermore, we derive a sufficient condition ensuring ergodicity and strict stationarity of the return process. Finally, we generalize the SDJ to a bivariate model where the intensity of the Poisson process follows a score-driven autoregression too. We conduct both an in-sample and an out-of-sample analysis focusing on the times series of the options written on the S&P500. The results obtained reveal that our score-driven approaches with jumps provide a very satisfactory agreement with empirical data and outperform existing GARCH models with jumps. To the best of our knowledge, score-driven models have not been used for derivative pricing so far.

C0672: **ESG dimensions in screening strategies: Impact on portfolio performance in periods of financial distress**

Presenter: **Beatrice Bertelli**, University of Modena and Reggio Emilia, Italy

Co-authors: Costanza Torricelli

In the last decades, professional investors have accelerated the integration of environmental, social and governance (ESG) dimensions into their investment decisions attracted by the opportunity to improve the risk-return performance. The aim, framed within the literature on socially responsible investments (SRI), is to analyse the impact of screening strategies based on ESG dimensions, especially in periods of financial distress such as the 2008 global recession and the 2020 Covid-19 pandemic. Socially responsible portfolios are built from 559 stocks that made up the EURO STOXX index from 2007 to 2021 by using both negative and positive screening strategies based on Bloomberg ESG disclosure scores and different screening thresholds. Hence socially responsible portfolios' Sharpe ratio and alpha are compared with a benchmark portfolio that represents a passive strategy. The main results suggest that ESG screens represent good-performing strategies in the long-term, whereas, when the observation period is narrowed down to times of financial distress, a broader and passive strategy appears to be better performing. Moreover, positive screening strategies, and in particular the ones that involve the social dimension, limit diversification benefits and are characterized by significant underperformance during periods of crises.

C0300: **Exploring risk hidden in syndicated loan networks: Evidence from real estate investment trusts**

Presenter: **Masayasu Kanno**, Nihon University, Japan

The aim is to assess interconnectedness and risk in the market for syndicated loans to Japan's real estate investment trusts (J-REITs) during the fiscal year 2013the first half of the fiscal year 2021. Network analysis indicates that Japanese major banks, large regional banks, and J-REITs play a central role in the network regarding degree centrality. Subsequently, the stress test investigates the resilience of a shock-propagation for a syndicated loan market. We found that no default contagion via a syndicated loan market is expected on the basis of syndicated loans outstanding at the end of 2021. Finally, we contribute to the literature regarding interconnectedness, credit risk, and systemic risk in the J-REIT syndicated loan network.

CC232 **Room Aula B** HIGH-DIMENSIONAL STATISTICS AND MODEL ASSESMENT **Chair: Germain Van Bever**

C0701: **Estimation in the high dimensional additive hazard model with l0 type of penalty**

Presenter: **Yunpeng Zhou**, The University of Hong Kong, Hong Kong

Co-authors: KC Yuen

High-dimensional data is commonly observed in survival data analysis. Penalized regression is widely applied for parameter selection given this type of data. The LASSO, SCAD and MCP methods are basic penalties developed in recent years in order to achieve a more accurate selection of parameters. The l0 penalty, which selects the best subset of parameters and provides unbiased estimation, is not fully researched due to its NP-hard complexity resulting from the non-smooth and non-convex objective function. Most methods developed so far focus on providing a smoothed version of the l0-norm which does not address the problem directly. Two Augmented Lagrangian-based algorithms are proposed for the additive hazard model, namely the ADMM-l0 method and the APM-l0 method, to approximate the optimal solution generated by the l0 penalty, among which the ADMM-l0 algorithm can achieve unbiased parameter estimation. Also, under moderate sample sizes, both methods perform well in selecting the best subset of parameters, especially in terms of controlling the false positive rate. The convergence of ADMM-l0 is proved under strict assumptions, and the performance of the proposed methods is illustrated using two DLBCL datasets.

C0693: **Bayesian group Lasso regression for genome-wide association studies**

Presenter: **Lanxin Li**, University of Glasgow, United Kingdom

Co-authors: Mayetri Gupta, Vincent Macaulay

Genome-wide association studies (GWAS) are designed to search across a genome-wide set of genetic variations (SNPs) from different individuals to find SNPs that are associated with a trait of interest. Many statistical methods for GWAS have limitations in accurately identifying SNPs underlying complex diseases (like heart disease), due to weak association signals from SNPs, local correlations between SNPs, and the sheer imbalance between the sizes of the available samples and candidate SNPs. We propose a Bayesian model framework, adapting ideas from Bayesian group Lasso regression, that clusters correlated SNPs into groups, and a population-based MCMC method to conduct powerful group selection in GWAS, to improve the accuracy and efficiency of detecting trait-associated regions. In this model, biological information relating to genomic structure and function can be used to elicit priors that improve the precision of SNP detection; signals from causative SNPs and SNPs correlated with causative ones can be accumulated to make the detection easier; and the total number of variables that need to be tested is vastly reduced. Results from a variety of contexts show that the proposed method improves on a variety of existing methods at association detection, especially when the signals coming from SNPs are weak.

C0703: **Statistical inference based on weighted divergence measures**

Presenter: **Vlad Barbu**, Universite de Rouen, France

Co-authors: Thomas Gkelsinis, Alexandros Karagrigoriou

The focus is on a class of hypotheses tests for goodness of fit and homogeneity between two samples. This type of test is constructed based on a particular type of discrepancy measure called weighted divergences. These measures allow us to focus on specific subsets of the support without, at the same time, losing the information of the others. With this method, we achieve a significantly more sensitive test than the classical ones, with comparable error rates. The appropriate asymptotic theory is presented according to Monte Carlo simulations for assessing the performance of the proposed test statistics.

CC224 Room Aula E LONGITUDINAL DATA **Chair: Silvia Pandolfi**

C0257: **Estimation of disease progression for ischemic heart disease using latent Markov with covariates**

Presenter: **Zarina Oflaz**, KTO Karatay University, Turkey

Contemporaneous monitoring of disease progression, in addition to early diagnosis, is important for the treatment of patients with chronic conditions. Chronic disease-related factors are not easily tractable, and the existing data sets do not clearly reflect them, making diagnosis difficult. The primary issue is that databases maintained by health care, insurance, or governmental organizations typically do not contain clinical information and instead focus on patient appointments and demographic profiles. Due to the lack of thorough information on potential risk factors for a single patient, investigations on the nature of disease are imprecise. We suggest the use of a latent Markov model with variables in a latent process because it enables the panel analysis of many forms of data. The purpose is to evaluate unobserved factors in ischemic heart disease (IHD) using longitudinal data from electronic health records. Based on the results we designate states as healthy, light, moderate, and severe to represent stages of disease progression. Gender, patient age, and hospital visit frequency are all significant factors in the development of the disease. Females acquire IHD more rapidly than males, frequently developing from moderate and severe disease. Additionally, it demonstrates that individuals under the age of 20 bypass the light state of IHD and proceed directly to the moderate state.

C0457: **Clustering with alignment and network inference to study the radiation response of endothelial cells**

Presenter: **Polina Arsenteva**, Institut de Mathematiques de Bourgogne, France

Co-authors: Vincent Paget, Olivier Guipaud, Fabien Milliat, Herve Cardot, Mohamed Amine Benadjaoud

Radiotherapy is a type of cancer treatment that may induce adverse effects on healthy tissues situated close to the irradiated tumor. It is important to study and compare different modes of radiotherapy in order to select those minimizing the potential undesirable consequences. The focus is on the response of endothelial cells, key actors in the appearance of radiation adverse effects. We study the expression of genes originating from transcriptomic in-vitro datasets that were collected for several time points under irradiated and non-irradiated conditions. The goal is to determine a small number of the most representative behavior types among all considered genes, and to identify potential biological pathways linked to the response to radiotherapy. The quantity of interest is radio-induced fold change: a measure of irradiation effect represented by the difference between the two experimental conditions over time. We propose a new approach based on modeling fold changes as random variables, and a new distance that allows accounting for uncertainties and correlations between variables. We designed a computationally efficient algorithm performing simultaneous clustering and alignment of fold changes' random estimators. Based on the obtained information, a gene network is inferred allowing to draw a comparison between different modes of radiotherapy.

C0559: **A personalized remote patient monitoring system using daily measurements of bodyweight, heart rate, and blood pressure**

Presenter: **Mehran Moazeni**, Utrecht Unveristy, Netherlands

Mortality rates and readmissions are prohibitively high for heart failure patients. These events are preceded by a period in which either one or a combination of bodyweight, heart rate, and blood pressure shift from a healthy baseline. This preceding shift offers an opportunity to early detect heart failure. Facilitating early detection of changes in biometric values, remote patient monitoring systems have been developed to record biometric values. Previously, simple algorithms were introduced to distinguish normal biometric observations from observations signalling heart failure by using absolute thresholds for all patients, rule-of-thumb and a moving average convergence-divergence algorithm. However, these algorithms have a poor performance in detecting heart failure as they display a high rate of false alarms. To alleviate this, we propose a novel personalized algorithm for two settings: single and combined biometric measurement monitoring. The algorithm is informed by cross-sectional and longitudinal data and uses a linear mixed-effect model to predict a personalised expected biometric value. Then, differences between the expected and observed value are used for a statistical process control chart, providing patient-specific thresholds for determining alarms. Comparing area-under-the-curve values showed that our personalised algorithm outperforms the above-mentioned algorithms for both settings. We discuss further improvements that could be incorporated to the algorithm

Friday 26.08.2022 15:00 - 16:00 Room: Aula 3 Chair: Ana Colubi Keynote talk 2

New graphical displays for classification

Speaker: **Peter Rousseeuw, KU Leuven, Belgium** Jakob Raymaekers

Classification is a major tool of statistics and machine learning. Several classifiers have interesting visualizations of their inner workings. We pursue a different goal, which is to visualize the cases being classified, either in training data or in test data. An important aspect is whether a case has been classified to its given class (label) or whether the classifier wants to assign it to a different class. This is reflected in the probability of the alternative class (PAC). A high PAC indicates label bias, i.e. the possibility that the case was mislabeled. The PAC is used to construct a silhouette plot which is similar in spirit to the silhouette plot for cluster analysis. The average silhouette width can be used to compare different classifications of the same dataset. We will also draw quasi-residual plots of the PAC versus a data feature, which may lead to more insight into the data. One of these data features is how far each case lies from its given class, yielding so-called class maps. The proposed displays are constructed for discriminant analysis, k-nearest neighbors, support vector machines, CART, random forests, and neural networks. The graphical displays are illustrated and interpreted on data sets containing images, mixed features, and texts.

Saturday 27.08.2022 13:55 - 14:45 Room: Aula 3 Chair: Maria Brigida Ferraro Keynote talk 1

A most surprising but useful result in semi-supervised learning (virtual)

Speaker: **Geoffrey McLachlan, University of Queensland, Australia** Daniel Ahfock

There has been increasing attention to semi-supervised learning approaches in machine learning to forming a classifier in situations where the training data for a classifier consists of a limited number of classified (labelled) observations but a much larger number of unclassified observations. This is because the procurement of classified data can be quite costly due to high acquisition costs and subsequent financial, time, and ethical issues that can arise in attempts to provide the true class labels for the unclassified data that have been acquired. The focus here is on the recent result that a classifier formed from a partially classified sample can actually have a smaller expected error rate than that if the sample were completely classified. This rather paradoxical outcome is able to be achieved by introducing a framework with a missingness mechanism for the missing labels of the unclassified observations. It is most relevant in commonly occurring situations in practice, where the unclassified data occur primarily in regions of relatively high entropy in the feature space thereby making it difficult for their class labels to be easily obtained.

Sunday 28.08.2022 11:35 - 12:30 Room: Aula 3 Chair: Erricos Kontoghiorghes Keynote talk 3

Distributed estimation through parallel approximants

Speaker: **Patrick Wolfe, Purdue University, United States** Aritra Chakravorty, William S Cleveland

Designing estimation algorithms that use the entirety of very large data sets is a core challenge in modern statistics. We provide a framework to address this challenge based on parallel approximants, which in turn yields scalable algorithms with accompanying consistency guarantees. Rather than employ approaches based on sampling, we instead first formalize the class of statistics which admit straightforward calculation in distributed environments through independent parallelization. We then show how to use such statistics to approximate arbitrary functional operators in appropriate spaces, yielding generic approximate inference procedures that do not require data to reside entirely in memory. We characterize the L^2 approximation properties of our approach, and discuss some canonical examples. A variety of avenues and extensions remain open for future work.

Friday 26.08.2022 16:30 - 18:10 Parallel Session B – SDS2022

SO012 Room Aula 3 RECENT ADVANCES IN DIMENSION REDUCTION AND RELATED METHODS Chair: Yuichi Mori

S0176: **Biplots in dimension reduction and clustering**

Presenter: **Michel van de Velden**, Erasmus University Rotterdam, Netherlands

Co-authors: Alfonso Iodice D Enza, Angelos Markos

In unsupervised learning, dimension reduction (e.g., PCA) and distance-based clustering are often applied sequentially: the distances used to cluster the observations are computed on the reduced dimensions. Since the dimension reduction step does not take into account the possible cluster structure, it is possibly detrimental to the clustering step. Methods for joint dimension reduction and clustering combine the two in a single optimization problem which is solved using iterative procedures alternating the two steps. Just like for principal component methods, different approaches have been proposed that deal with continuous, categorical or mixed-type data. In particular, for continuous data, reduced K-means combines principal component analysis with K-means clustering; for categorical data, cluster correspondence analysis combines correspondence analysis with K-means; for mixed-type data, mixed Reduced K-means combines factor analysis for mixed data with K-means. The biplot visualization of the solution is of particular interest for interpretation purposes: the low-dimensional map can be very helpful for cluster characterization. We illustrate the use of biplots in the context of dimension reduction and clustering.

S0182: **Asymmetric MDS with sparse regularization term**

Presenter: **Kensuke Tanioka**, Doshisha University, Japan

Co-authors: Hiroshi Yadohisa

Asymmetric (dis)similarity data are (dis)similarity data in which the relationship between the proximity from subject i to subject j and the proximity from subject j to subject i is not necessarily equivalent and can be observed in various fields. Asymmetric multidimensional scaling (Asymmetric MDS) is one of the methods to visualize asymmetric relationships among objects from such asymmetric (dis)similarity data. Although various asymmetric MDS methods have been proposed, we assume a situation in which covariate data are given as external information for the same subject as the asymmetric (dis)similarity data. In concretely, we propose an asymmetric multidimensional scaling method that can draw asymmetric relationships between asymmetric results and objects that can be visually interpreted, and identify covariates of external information that are related to the asymmetry. In the proposed method, coordinates and parameters in a low-dimensional space are represented by a linear combination of covariate data and a loading matrix, and by introducing a regularization term, such as lasso, it is possible to simultaneously implement variable selection to interpret asymmetric relationships.

S0187: **Regularized functional subspace clustering**

Presenter: **Yoshikazu Terada**, Osaka University; RIKEN, Japan

Co-authors: Michio Yamamoto

The intrinsic high dimensional nature of functional data often makes a possible good performance in supervised classification for functional data. Using the projection into the finite-dimensional subspace, we can extract the intrinsic high-dimensional nature from functional data. Several subspace clustering methods are proposed for functional data. However, the projected data do not necessarily reflect the hidden cluster structure in some of these methods. We propose a new regularized subspace clustering algorithm for functional data. We can ensure that the objective function monotonically decreases at each iteration for the proposed algorithm. Moreover, we study the asymptotic properties of the proposed clustering algorithm. Finally, we demonstrate that the proposed method provides better performance than the existing clustering methods for both simulated and real data through numerical experiments.

S0153: **Estimation and inference of A heteroskedasticity model with latent semiparametric factors for panel data analysis**

Presenter: **Wen Zhou**, Colorado State University, United States

Co-authors: Lyuou Zhang, Haonan Wang

Estimation and inference of a flexible subject-specific heteroskedasticity model for analyzing large scale panel data is considered. The model employs latent semiparametric factor structure to simultaneously account for the heteroskedasticity across subjects and contemporaneous correlations. Specifically, the heteroskedasticity across subjects is modeled by the product of unobserved stationary process of factors and subject-specific covariate effect. Serving as the loading, the covariate effect is further modeled through the additive model. We propose a two-step procedure for estimation. First, the latent factor process and nonparametric loading are estimated via projection-based methods. The estimation of regression coefficients is further conducted through the generalized least squares type approach. Theoretical validity of the two-step procedure is carefully documented. By scrupulously examining the non-asymptotic rates for recovering the latent factor process and its loading, we further study the properties of the estimated regression coefficients. In particular, we establish the asymptotic normality of the proposed two-step estimate of regression coefficients. The proposed regression coefficient estimator is also shown to be asymptotically efficient. This leads to a more efficient confidence set of the regression coefficients.

SO015 Room Aula 4 BAYESIAN LEARNING Chair: Igor Pruenster

S0181: **Statistical guarantees for variational automatic relevance determination**

Presenter: **Feng Liang**, University of Illinois at Urbana-Champaign, United States

Co-authors: Zihe Liu

The Automatic Relevance Determination (ARD) model is studied for high-dimensional linear regression under sparsity constraints. For each regression parameter, the ARD prior places a Gaussian distribution with mean zero, and variance is a hyper-parameter that needs to be learned from the data. We focus on a variational procedure, which approximates the posterior distribution by independent Gaussian distributions, one for each parameter. It can be shown that for some parameters the corresponding Gaussian distribution will degenerate to a point mass at zero; that is, some variables will be automatically filtered out by the ARD procedure. Although ARD and this variational framework have been studied before, little is known about the theoretical properties of the variational solution. The main contribution is to establish convergence results, in terms of parameter estimation and variable selection, for the variational solution of an ARD model.

S0175: **An application of square root transformation for optimal prior selection**

Presenter: **Cristiano Villa**, Newcastle University, United Kingdom

Co-authors: Stephen Walker, Alfred Kume

The pooling of opinions is a big area of research and has been for a number of decades. The idea is to obtain a single belief probability distribution from a set of expert opinion belief distributions. A new way is provided to provide a resultant prior opinion based on the optimal information among all possible linear combinations of the prior densities, including negative components. This is done in the square-root density space which is identified with the positive orthant of the Hilbert unit sphere of differentiable functions. It can be shown that the optimal prior is easily identified as an extrinsic mean in the sphere. For distributions belonging to the exponential family, the resulting calculations do not require numerical integration and can be immediately implemented in the Bayesian analysis. The idea can also be adopted for any neighbourhood of a chosen base prior and spanned by a finite set of "contaminating" directions.

S0185: **Bayesian nonparametric analysis of calcium imaging data**
Presenter: **Antonio Canale**, University of Padua, Italy
Co-authors: Michele Guindani, Laura D Angelo

Recent advancements in miniaturized fluorescence microscopy have made it possible to investigate neuronal responses to external stimuli in awake behaving animals through the analysis of intracellular calcium signals. We will discuss several challenges that this novel and complex type of data pose and how they can be solved by means of flexible Bayesian nonparametric models. The proposed solutions exploit several recent advances in Bayesian nonparametric including nonparametric dependent mixture priors to borrow information between experiments and discover similarities in the distributional patterns of neuronal responses and inner spike-and-slab nonparametric models to jointly model different patterns of neuronal activity or the lack of thereof.

S0177: **A conjugate prior for the Dirichlet process precision parameter**
Presenter: **Tommaso Rigon**, University of Milano-Bicocca, Italy
Co-authors: Alessandro Zito, David Dunson

A new prior distribution is presented for the precision parameter of a Dirichlet process. We show how this prior is conjugate to the distribution of the number of distinct values arising from the process. Moments, properties and hyperparameters interpretation of the distribution are extensively studied, as well as its relationship with the class of exponential families. Interestingly, certain choices for the hyperparameters allow to compute the normalizing constant explicitly. We show how this allows to draw a parallel with common Bayesian nonparametric models within the class of Gibbs-type processes. We illustrate practical advantages of using this prior over common alternatives proposed in the literature when adopting a Dirichlet process-based clustering algorithm.

Saturday 27.08.2022 09:00 - 10:15 Parallel Session C – SDS2022

SO008 Room Aula 3 STATISTICAL LEARNING FOR NETWORK DATA WITH APPLICATIONS Chair: Yousri Slaoui

S0172: **Mixture of longitudinal factor analysis for modelling heterogeneous longitudinal multivariate data**
Presenter: **Amine Ounajim**, University of Poitiers, France
Co-authors: Yousri Slaoui, Pierre-Yves Louis, Denis Frasca, Philippe Rigoard

In order to study the evolution of several observed outcomes among patients, it is important to focus on longitudinal trends among latent variables using joint modeling based on covariance structures between these observed outcomes. However, this type of data might represent heterogeneity over time and among groups of individuals. To address this problem, some authors have proposed factor analyzer mixture models, which estimate different factor loadings for different subpopulations, which are represented by a latent class variable. We propose here an extension to the factor analysis framework where group non-invariance is taken into account using a mixture model. We start by defining the mixture of the longitudinal factor analysis model and its parameters. Then, we propose an EM algorithm to estimate the model. We also develop a Bayesian information criterion to identify the number of components in the mixture. We then discuss the comparability of the latent factors obtained between subjects in different latent groups. Finally, we apply the model to simulated and real data of patients with postoperative chronic pain.

S0204: **Usage of Bayesian neural network in deep reinforcement learning**
Presenter: **Leo Grill**, Universite de Poitiers, France
Co-authors: Yousri Slaoui, David Nortershauser, Stephane Le Masson

While reinforcement learning finds applications in various domains, the Bayesian methods have special attention, especially deep learning. They can be used to deal with issues such as overfitting or explainable modeling. We developed the usage of using a Bayesian neural network as an actor in the actor-critic algorithms. We show the benefits of using these methods in deep-reinforcement learning.

S0202: **Extension of the stochastic block model to handle networks with weighted nodes with application to EEG data**
Presenter: **Yousri Slaoui**, University of Poitiers, France

The focus is on the analysis of weighted networks, finite graphs where each edge is associated with a weight representing the intensity of its strength. We introduce an extension of the binary stochastic block model (SBM), called binomial stochastic block model (bSBM). This question is motivated by the study of co-citation networks in a text mining context where the data is represented by a graph. The nodes are words and each edge joining two words is weighted by the number of documents included in the corpus simultaneously citing that pair of words. We develop an inference method based on the variational expectation maximization (VEM) algorithm and another based on the Bayesian variational expectation maximization (BVEM) algorithm to estimate the parameters of the proposed model as well as to classify the words of the network. Then we adopt a method based on the maximization of an integrated classification likelihood (ICL) criterion to select the optimal model and the number of clusters. Applications to real data are adopted to show the effectiveness of both methods as well as to compare them. Finally, we develop a multi-attribute SBM to handle networks with weights associated with nodes. We motivate this method with an application that aims at developing a tool to help specify different cognitive processes performed by the brain during writing preparation.

S0203: **Multivariate extremes for risk assessment**
Presenter: **Salah El Adlouni**, Universite de Moncton, Canada

Multivariate characterisation of extremes is a reference decision tool for hydrometeorological risk. It allows estimating extremes and their return periods. Generally, frequency curves are estimated separately through a univariate model for each parameter. Multivariate risk is then estimated based on independent variables. This hypothesis is very strong and is not necessarily verified for several hydroclimatic variables. The aim is to examine the effects of the independence hypothesis by proposing a multivariate model that considers the dependencies between the variables. The multivariate model is based on D-vine copulas to explore the intra-covariate dependencies structures. An illustration of the proposed model is presented for the sizing of pipes and hydrological retention basins in flooding events.

SO031 Room Aula 4 MODELS FOR THE ANALYSIS AND CLASSIFICATION OF HETEROGENEOUS DATA Chair: Geoffrey McLachlan

S0163: **Semiparametric finite mixture of regression models with Bayesian P-splines**
Presenter: **Marco Berrettini**, University of Bologna, Italy
Co-authors: Giuliano Galimberti, Saverio Ranciati

Mixture models provide a useful tool to account for unobserved heterogeneity and are at the basis of many model-based clustering methods. To gain additional flexibility, some model parameters can be expressed as functions of concomitant covariates. A semiparametric finite mixture of regression models is defined, with concomitant information assumed to influence both the component weights and the conditional means. The proposed contribution replaces the linear predictors with a smooth function of the covariate considered. An estimation procedure within the Bayesian paradigm is suggested, where the smoothness of the covariate effects is controlled by suitable choices for the prior distributions of the spline coefficients. A data augmentation scheme based on different random utility models is exploited to describe the mixture weights as functions of the covariate. The performance of the proposed methodology is investigated via simulation experiments, and applications to real datasets are discussed.

S0191: **An EM algorithm for semi-supervised learning with data augmentation**
Presenter: **Daniel Ahfock**, University of Queensland, Australia
Co-authors: Geoffrey McLachlan

A popular strategy for semi-supervised learning is to use unlabelled data to construct a regularization term to guide the training of a classification model. Data augmentation is a technique for the generation of artificial unlabelled data through the perturbation of available data. Data augmentation is frequently combined with consistency regularization, whereby the model is encouraged to make similar predictions on the original and perturbed data. We propose a consistency regularization technique based on the Bhattacharyya coefficient for use with data augmentation. An EM algorithm is developed for the maximization of the regularized likelihood. The asymptotic variance of the regularized semi-supervised estimator can be linked to the asymptotic variance of an unregularized supervised estimator given a completely classified sample. Theoretical analysis suggests that semi-supervised learning can be competitive with fully supervised learning under assumptions on the quality of the data augmentation procedure.

S0192: **$(\text{SAM})^2$: A family of sequential sample averaging algorithms via majorization–minimization**
Presenter: **Hien Nguyen**, University of Queensland, Australia

In practice, many data sets are retrieved sequentially or are too large to be stored in the memory of typical machines available to most analysts. We consider a sequential sampling approach for constructing optimization algorithms for many machine learning and statistical settings that can alleviate the difficulties caused by the aforementioned issues. The algorithms combine a sequential sampling scheme with a majorization–minimization approach to producing a globally convergent family of algorithms that has broad applicability.

Saturday 27.08.2022 10:45 - 12:25 Parallel Session D – SDS2022

SO010 Room Aula 3 TEXT BASED INDICATORS IN ECONOMICS AND FINANCE **Chair: Peter Winker**

S0180: **Difference in SDG reporting of research articles using zero-shot text classification**
Presenter: **Christoph Funk**, Justus-Liebig-University Giessen, Germany
Co-authors: Elena Toenjes, Lutz Breuer, Ramona Teuber

In September 2015, the United Nations (UN) set an agenda to transform our world by 2030 with the adoption of 17 Sustainable Development Goals (SDGs), 169 targets and 231 indicators for monitoring progress towards the goals. Since then, the academic literature on the SDGs has grown tremendously. The analysis of such large amounts of textual data requires the use of Natural language processing (NLP) techniques. Here, we apply zero-shot classification as a text mining tool on SDG-related scientific articles to analyze the scientific discourse on the 17 SDGs. The contributions are four-fold. First, we review the scientific literature on the SDGs in order to draw conclusions about the scientific discourses worldwide. Second, we show that abstracts contain the most relevant information from scientific articles related to the discussed SDGs. This means that applying NLP techniques on abstracts instead of the whole article is sufficient, which in turn saves computational power and thus time. Third, we show that zero-shot text classification can be a useful tool to label extensive textual information and thus might be relevant for policymakers by providing information beyond the typical UN indicators in an efficient manner. Fourth, we compare the scientific discourse with the official average UN SDG indicator scores.

S0196: **A novel fuzzy spectral clustering approach for text data**
Presenter: **Irene Cozzolino**, Universita La Sapienza, Italy
Co-authors: Maria Brigida Ferraro, Peter Winker

Spectral clustering methodologies have been widely used in text classification tasks due to their good performances and solid theoretic foundations which do not assume any prearranged structure in the data. However, very few contributions have been proposed for unsupervised classification techniques. We focus on the employment of the spectral clustering algorithm when analysing unlabelled text documents. A crucial point with this method consists in the construction of an adequate similarity matrix to use as input of the algorithm. This aspect has motivated us to introduce a new similarity measure for text data based on a weighted combination of both sequence and set similarities, in order to also capture the inherent sequential nature of text files that can be seen as an ordered sequence of words. Furthermore, a new fuzzy version of spectral clustering has been introduced to use in combination with the proposed similarity. The newly introduced document clustering algorithm has been evaluated by means of benchmark and real data sets.

S0156: **Anti-pandemic restrictions, uncertainty and sentiment in seven countries**
Presenter: **Svetlana Makarova**, University College London, United Kingdom
Co-authors: Wojciech Charemza, Krzysztof Rybinski

The purpose is to evaluate how the stringency of government anti-pandemic policy might affect economic policy uncertainty in countries with different degrees of press freedom and various reporting styles and writing conventions. We apply a text-based measure of uncertainty using data from over 400,000 press articles from Belarus, Kazakhstan, Poland, Russia, Ukraine, the UK, and the US published during the waves of the Covid-19 pandemic before the wide-scale vaccination programmes were introduced. The measure accounts for pandemic-related words, and sentiment scores to weight the selected articles. We find that the level of anti-pandemic stringency negatively affects uncertainty, while a change in that level has the opposite effect.

S0171: **Text based innovation indicators a progress report**
Presenter: **Peter Winker**, University of Giessen, Germany
Co-authors: David Lenz, Albina Latifi

There exist many indicators for innovative activity. The projects TOBI and DynTOBI aim at developing novel text-based indicators based on the information provided in websites of firms and on news articles from a technology-related online news provider. The presentation will focus on the latter dataset, which allows describing innovation diffusion over time. It is described which steps are required to transform the raw textual data into time series which might reflect the diffusion of new products or technologies. While the presented results will be mainly explorative, the approach might be developed further for prediction purposes. The first step of the analysis consists in applying computational methods from natural language processing to identify latent topics in the text corpus and to obtain associated time series of topic weights. Furthermore, a labeling of innovation topics is performed by experts. In a second step, methods from functional data analysis (FDA) are applied to categorize these time series in clusters. For this purpose, an implementation of the global search heuristic Threshold Accepting (TA) is applied, which appears provides better and more robust results compared to standard sequential techniques such as k-means. The identified clusters of prototypical innovation diffusion trends show some variability as compared to the standard textbook shape. Moreover, the approach allows to uncover different stages of innovation diffusion.

SO023 Room Aula 4 COMPUTATIONAL AND METHODOLOGICAL CHALLENGES IN ENVIRONMENTAL DATA **Chair: Abhirup Datta**

S0152: **Bayesian model selection for ultrahigh dimensional doubly intractable distributions**
Presenter: **Jaewoo Park**, Yonsei University, Korea, South
Co-authors: Ick Hoon Jin

Doubly intractable distributions commonly arise in many complex statistical models in physics, epidemiology, ecology, social science, among other disciplines. With an increasing number of model parameters, they often result in ultrahigh dimensional posterior distributions; this is a challenging problem and is crucial for developing the computationally feasible approach. A particularly important application of ultrahigh dimensional doubly intractable models is network psychometrics, which gets attention in item response analysis. However, its parameter estimation method, maximum pseudo-likelihood estimator (MPLE) combining with lasso certainly ignores the dependent structure, so that it is inaccurate. To tackle this problem, we propose a novel Markov chain Monte Carlo methods by using Bayesian variable selection methods to identify strong interactions automatically. With our new algorithm, we address some inferential and computational challenges: (1) likelihood functions involve doubly-intractable normalizing functions, and (2) increasing number of items can lead to ultrahigh dimensionality in the model. We illustrate the application of our approaches to challenging simulated and real item response data examples for which studying local dependence is very difficult. The proposed algorithm shows significant inferential gains over existing methods in the presence of strong dependence among items.

S0160: **Computer model calibration with time series data using deep learning and quantile regression**
Presenter: **Won Chang**, University of Cincinnati, United States
Co-authors: Jiali Wang, Saumya Bhatnagar, Seonjin Kim

Computer models play a key role in many scientific and engineering problems. One major source of uncertainty in computer model experiments is input parameter uncertainty. Computer model calibration is a formal statistical procedure to infer input parameters by combining information from model runs and observational data. The existing standard calibration framework suffers from inferential issues when the model output and observational data are high-dimensional dependent data, such as large time series, due to the difficulty in building an emulator and the non-identifiability between effects from input parameters and data-model discrepancy. To overcome these challenges, we propose a new calibration framework based on a deep neural network (DNN) with long short-term memory layers that directly emulates the inverse relationship between the model output and input parameters. Adopting the learning with noise idea, we train our DNN model to filter out the effects of data-model discrepancy on input parameter inference. We also formulate a new way to construct interval predictions for DNN using quantile regression to quantify the uncertainty in input parameter estimates. Through a simulation study and real data application with the Weather Research and Forecasting Model Hydrological modeling system (WRF-Hydro), we show our approach can yield accurate point estimates and well-calibrated interval estimates for input parameters.

S0162: **Nearest neighbors processes for non-Gaussian geostatistical data**

Presenter: **Bruno Sanso**, University of California Santa Cruz, United States

Co-authors: Xiaotian Zheng, Thanasis Kottas

A framework is presented for non-Gaussian spatial processes that encompasses large distribution families. Spatial dependence for a set of irregularly scattered locations is described with a mixture of pairwise kernels. Focusing on the nearest neighbors of a given location, within a reference set, we obtain a valid spatial process: the nearest neighbor mixture transition distribution process (NNMP). We develop conditions to construct general NNMP models with arbitrary pre-specified marginal distributions. Essentially, NNMPs are specified by a bi-variate distribution, with suitable marginals, used to specify the mixture transition kernels. Such a distribution can be spatially varying, to capture non-homogeneous spatial features. The mixture structure of the model allows for efficient MCMC-based exploration of the posterior distribution of the model parameters, even for a relatively large number of locations. We illustrate the capabilities of NNMPs with observations corresponding to distributions with different non-Gaussiancharacteristics: Long tails; Compact support; skewness. We extend NNMPs to tackle discrete-valued distributions using a continuous extension for the discrete bivariate copulas to enhance computational efficiency and stability. We illustrate the discrete NNMP with data corresponding to counts from the North American Bird Survey.

S0184: **Long memory random fields on regular lattices**

Presenter: **Luigi Ippoliti**, University of Chieti Pescara, Italy

Co-authors: Angela Ferretti, Rajendra Bhansali, Pasquale Valentini

The motivation is drawn from applications in environmental sciences where empirical evidence of slow decay of correlations has been found for data observed on a regular lattice. Different types of models have been proposed for analysing such data. We introduce the class of conditional autoregressive fractional integrated moving average (CARFIMA) models as a suitable framework for studying environmental data with long-range spatial correlation. For this class, we provide detailed descriptions of some representative models, make the necessary comparison with some other existing models, and discuss some important inferential and computational issues on estimation, simulation and long memory process approximation. Results from model fit comparison and predictive performance of CARFIMA models are discussed through a statistical analysis of satellite land surface temperature data.

Saturday 27.08.2022 14:55 - 16:35 Parallel Session F – SDS2022

SO021 Room Aula 3 ANALYTICAL CHALLENGES WITH COMPLEX DATA ANALYSIS **Chair: Anna Gottard**

S0161: **Bayesian inference in large complex networks**

Presenter: **Snigdhansu Chatterjee**, University Of Minnesota, United States

Datasets are considered where the observations are over vertices of large graphs or networks, and there may be high-dimensional features associated with each vertex. For such datasets, the nature of the observations may depend not only on the features of the vertex they are associated with, but also on features of other vertices depending on the properties of the edges. We discuss Bayesian inference in models for such data. We consider generalizations of traditional network models like the stochastic block model, random dot product model, and so on, and propose different computational strategies for obtaining posterior distributions of interest. Apart from the traditional Markov Chain Monte Carlo approach, we study approximate Bayesian computations and some recently proposed piecewise deterministic Monte Carlo approaches. Comparisons between the different algorithms based on their computational efficiencies and statistical accuracies are discussed. Applications in several real data problems are discussed.

S0198: **Perturbing data to address dataset shift in supervised classification**

Presenter: **Angela Montanari**, Alma mater studiorum-Universita di Bologna, Italy

Co-authors: Laura Anderlucci

In supervised classification, dataset shift occurs when for the units in the test set a change in the distribution of a single feature, a combination of features, or the class boundaries, is observed with respect to the training set. As a result, in real data applications, the common assumption that the training and testing data follow the same distribution is often violated. Dataset shift might be due to several reasons; the focus is on what is called "covariate shift", namely the conditional probability $p(y|x)$ remains unchanged, but the input distribution $p(x)$ differs from training to test set. Random perturbation of variables or units when building the classifier can help in addressing this issue. Evidence of the performance of the proposed approach is obtained on simulated and real data.

S0205: **Distribution-invariant differential privacy**

Presenter: **Xuan Bi**, University of Minnesota, United States

Co-authors: Xiaotong Shen

Differential privacy is becoming one gold standard for protecting the privacy of publicly shared data. It has been widely used in social science, data science, public health, information technology, and the U.S. decennial census. Nevertheless, to guarantee differential privacy, existing methods may unavoidably alter the conclusion of the original data analysis, as privatization often changes the sample distribution. This phenomenon is known as the trade-off between privacy protection and statistical accuracy. We mitigate this trade-off by developing a distribution-invariant privatization (DIP) method to reconcile both high statistical accuracy and strict differential privacy. As a result, any downstream statistical or machine learning task yields essentially the same conclusion as if one used the original data. Numerically, under the same strictness of privacy protection, DIP achieves superior statistical accuracy in two simulations and on three real-world benchmarks.

S0206: **Bayesian networks for dihedral angles**

Presenter: **Anna Gottard**, University of Firenze, Italy

Co-authors: Agnese Panzera

A crucial topic in structural bioinformatics is predicting the three-dimensional structure of a protein, as determined by dihedral angles. It is believed that the amino acid sequence of a protein, its primary structure, incorporates the information needed to determine its shape, in turn governing the biological activity. For some proteins, the secondary structure and the functionality may vary with the membrane composition of the peptide habitat. A step forward in protein prediction could be understanding the conditional independence structure linking the angles of a protein. Graphical models are a well-known tool for analysing conditional independence between random variables. Dihedral angles are a special kind of random variable called a circular variable. The intrinsic characteristics of this kind of variable require ad hoc distributions. We explore possible specifications of Bayesian Networks for dihedral angles, assuming that the primary structure provides a natural ordering of the nodes. We analyse alternative parameterisations of Bayesian Networks for Conditional von Mises distribution and Inverse Stereographic distribution. We also provide possible inferential procedures for estimation and graph learning. As an illustration, we apply the proposals to the Methionine data.

SO033 Room Aula 4 NONASYMPTOTIC STATISTICS AND ECONOMETRIC **Chair: Zhao Ren**

S0168: **Heteroskedastic sparse PCA in high dimensions**

Presenter: **Zhao Ren**, University of Pittsburgh, United States

Principal component analysis (PCA) is one of the most commonly used techniques for dimension reduction and feature extraction. Though it has been well-studied for high-dimensional sparse PCA, little is known when the noise is heteroskedastic, which turns out to be ubiquitous in many scenarios, like biological sequencing data and information network data. We propose an iterative algorithm for sparse PCA in the presence of heteroskedastic noise, which alternatively updates the estimates of the sparse eigenvectors using the power method with adaptive thresholding in one step, and imputes the diagonal values of the sample covariance matrix to reduce the estimation bias due to heteroskedasticity in the other step. Our procedure is computationally fast and provably optimal under the generalized spiked covariance model, assuming the leading eigenvectors are sparse. A comprehensive simulation study demonstrates its robustness and effectiveness under various settings.

S0170: **L^2 inference of change-points for high-dimensional time series**

Presenter: **Weining Wang**, University of York, United Kingdom

A new inference method is proposed for multiple change-point detections of high-dimensional time series. The proposed approach targets dense or spatially clustered cross-sectional signals. An L^2-aggregated statistics is adopted in the cross-sectional dimension to detect multiple mean shifts for high-dimensional dependent data, and then followed by maximum over time. On the theory front, we develop the asymptotic theory concerning the limiting distributions of the change-point test statistics under both the null and alternatives, and we establish the consistency of the estimated break dates. The core of our theory is to extend a high-dimensional Gaussian approximation theorem to non-stationary dependent data, in particular for an $L^2 - L^\infty$ type statistics which is not available in the literature. Moreover, to facilitate the inference of breaks with natural clusters in the cross-sectional dimension, we also provide asymptotic properties of the test statistics with spatial dependence. Numerical simulations demonstrate the power enhancement of our newly proposed testing method relative to other existing techniques.

S0173: **Kronecker product approximation for matrix approximation, denoising and completion**

Presenter: **Rong Chen**, Rutgers University, United States

The problem of matrix approximation, denoising and completion induced by the Kronecker product decomposition is considered. Specifically, we propose to approximate a given matrix by the sum of a few Kronecker products of smaller matrices, which we refer to as the Kronecker product approximation (KoPA). Because the Kronecker product is an extension of the outer product from vectors to matrices, KoPA extends the low-rank matrix approximation and includes the latter as a special case. Compared with the latter, KoPA also offers greater flexibility, since it allows the user to choose the configuration, which are the dimensions of the two matrices forming the Kronecker product. As the configuration to be used is usually unknown, an extended information criterion is used to select the configuration. The model is extended to allow for multiple terms with different configurations (hybrid-KoPA) for more efficient approximation and denoising. It is also used for matrix completion tasks, with superior theoretical and numerical properties.

S0174: **Robust estimation and inference for expected shortfall regression**

Presenter: **Wenxin Zhou**, University of California San Diego, United States

Co-authors: Kean Ming Tan, Xuming He

Expected Shortfall (ES), also known as superquantile or conditional Value-at-Risk, has been recognized as an important measure in risk analysis and stochastic optimization, and is also finding applications beyond these areas. In finance, it refers to the conditional expected return of an asset given that the return is below some quantile of its distribution, namely its Value-at-Risk (VaR). We consider a recently proposed joint regression framework that simultaneously models the quantile and the ES of a response variable given a set of covariates, for which the state-of-the-art approach is based on minimizing a joint loss function that is non-differentiable and non-convex. This inevitably raises numerical instabilities, and thus limits its applicability for analyzing large-scale data. Motivated by the idea of using Neyman-orthogonal scores to reduce sensitivity with respect to nuisance parameters, we propose a statistically robust (to heavy-tailed data) and computationally efficient two-step procedure for fitting joint quantile and ES regression models. Under increasing-dimensional settings, we establish explicit non-asymptotic bounds on estimation and Gaussian approximation errors, which lay the foundation for statistical inference of ES regression. Numerical studies demonstrate the superior statistical performance and numerical efficiency of the proposed method.

Saturday 27.08.2022 17:05 - 18:45 Parallel Session G – SDS2022

SO006 Room Aula 3 FUNCTIONAL AND OBJECT DATA ANALYSIS **Chair: Xiaoke Zhang**

S0166: **Modeling time-varying random objects and dynamic networks**
Presenter: **Paromita Dubey**, University of Southern California, United States
Co-authors: Hans-Georg Mueller

Samples of dynamic or time-varying networks and other random object data such as time-varying probability distributions are increasingly encountered in modern data analysis. Common methods for time-varying data such as functional data analysis are infeasible when observations are time courses of networks or other complex non-Euclidean random objects that are elements of general metric spaces. In such spaces, only pairwise distances between the data objects are available. We combat this complexity by a generalized notion of mean trajectory taking values in the object space. For this, we adopt pointwise Frechet means and then construct pointwise distance trajectories between the individual time courses and the estimated Frechet mean trajectory, thus representing the time-varying objects and networks by functional data. Functional principal component analysis of these distance trajectories can reveal interesting features of dynamic networks and object time courses and is useful for downstream analysis. We demonstrate desirable asymptotic properties of sample-based estimators for suitable population targets under mild assumptions. The utility of the proposed methodology is illustrated with dynamic networks, time-varying distribution data and longitudinal growth data.

S0167: **Learning the regularity of curves in functional data analysis and applications**
Presenter: **Steven Golovkine**, University of Limerick, Ireland
Co-authors: Valentin Patilea, Nicolas Klutchnikoff

Combining information both within and across trajectories, we propose a simple estimator for the local regularity of the trajectories of a stochastic process. Independent trajectories are measured with errors at randomly sampled time points. The proposed approach is model-free and applies to a large class of stochastic processes. Non-asymptotic bounds for the concentration of the estimator are derived. Given the estimate of the local regularity, we build a nearly optimal local polynomial smoother from the curves from a new, possibly very large sample of noisy trajectories. We derive non-asymptotic pointwise risk bounds uniformly over the new set of curves. Our estimates perform well in simulations, in both cases of differentiable or non-differentiable trajectories.

S0164: **Spherical autoregressive models, with application to distributional and compositional time series**
Presenter: **Changbo Zhu**, University of California, Davis, United States
Co-authors: Hans-Georg Mueller

A new class of autoregressive models is introduced for spherical time series, where the dimension of the spheres on which the observations of the time series are situated may be finite-dimensional or infinite-dimensional as in the case of a general Hilbert sphere. Spherical time series arise in various settings. We focus on distributional and compositional time series. Applying a square root transformation to the densities of the observations of a distributional time series maps the distributional observations to the Hilbert sphere, equipped with the Fisher-Rao metric. The challenge in modeling such time series lies in the intrinsic non-linearity of spheres and Hilbert spheres. To address this difficulty, we consider rotation operators to map observations on the sphere. Specifically, we introduce a class of skew-symmetric operators such that the associated exponential operators are rotation operators that for each given pair of points on the sphere map one of the points to the other one. We exploit the fact that the space of skew-symmetric operators is Hilbertian to develop autoregressive modeling of geometric differences that correspond to rotations of spherical and distributional time series. We showcase the models with a time series of yearly observations of bivariate distributions of the minimum/maximum temperatures for a period of 120 days during each summer for the years 1990-2018 at Los Angeles (LAX) and John F. Kennedy (JFK) international airports.

S0159: **Proximal learning for individualized treatment regimes under unmeasured confounding**
Presenter: **Xiaoke Zhang**, George Washington University, United States
Co-authors: Zhengling Qi, Rui Miao

Data-driven individualized decision-making has recently received increasing research interest. Most existing methods rely on the assumption of no unmeasured confounding, which unfortunately cannot be ensured in practice, especially in observational studies. Motivated by the recently proposed proximal causal inference, we develop several proximal learning approaches to estimating optimal individualized treatment regimes (ITRs) in the presence of unmeasured confounding. In particular, we establish several identification results for different classes of ITRs, exhibiting the trade-off between the risk of making untestable assumptions and the value function improvement in decision making. Based on these results, we propose several classification-based approaches to finding a variety of restricted in-class optimal ITRs and developing their theoretical properties. The appealing numerical performance of our proposed methods is demonstrated via an extensive simulation study and a real data application.

SO019 Room Aula 4 MACHINE LEARNING FOR SPATIAL ANALYSIS **Chair: Abhirup Datta**

S0183: **Generalizing random forests for spatially correlated data**
Presenter: **Abhirup Datta**, Johns Hopkins Bloomberg School of Public Health, United States

Spatial linear mixed-models, consisting of a linear covariate effect and a Gaussian process (GP) distributed spatial random effect, are widely used for analyses of geospatial data. We consider the setting where the covariate effect is nonlinear. Random forests (RF) are popular for estimating nonlinear regression functions but applications of RF for spatial data have often ignored the spatial correlation. We show that this impacts the performance of RF adversely. We propose RF-GLS, a novel and well-principled and parsimonious extension of RF, for estimating nonlinear covariate effects in spatial mixed models where the spatial correlation is modeled using GP. RF-GLS extends RF in the same way generalized least squares (GLS) fundamentally extends ordinary least squares (OLS) to accommodate for dependence in linear models. RF becomes a special case of RF-GLS, and is substantially outperformed by RF-GLS for both estimation and prediction across extensive numerical experiments with spatially correlated data. RF-GLS can be used for functional estimation in other types of dependent data like time series. We also provide, to our knowledge, the first asymptotic consistency results for tree and forest estimators under spatial dependence.

S0190: **Data Compression and Distributed Inference in Spatial Analysis**
Presenter: **Rajarshi Guhaniyogi**, Texas A & M university, United States

Bayesian data sketching for spatial regression models is introduced to obviate computational challenges presented by large numbers of spatial locations. To address the challenges of analyzing very large spatial data, we compress spatially oriented data by a random linear transformation to achieve dimension reduction and conduct inference on the compressed data. Our approach distinguishes itself from several existing methods for analyzing large spatial data in that it requires neither the development of new models or algorithms nor any specialized computational hardware while delivering fully model-based Bayesian inference. Well-established methods and algorithms for spatial regression models can be applied to compressed data. We further extend data sketching idea to offer important advantages in the realm of distributed Bayesian inference.

S0208: **Classification of ENSO phases using topological data analysis**
Presenter: **Adam Jaeger**, Wichita State University, United States

The El Nino Southern Oscillation (ENSO) is one of the most powerful climate phenomena that can change global air circulation, affecting temperature and rainfall around the planet. Classification of these phases has traditionally relied on average sea surface temperatures in the equatorial Pacific without consideration of any structural information. Topological data analysis(TDA) is an innovative approach which focuses on a data set's "shape" or topological structures such as loops, holes, and voids. We used TDA to describe the homology groups of the two-dimensional function determined by sea surface temperatures of the tropical Pacific Ocean and utilize these summaries as a potential alternative to the prediction of ENSO phase.

S0201: **Nonstationary spatial modeling of massive global satellite data**

Presenter: **Dorit Hammerling**, Colorado School of Mines, United States

Co-authors: Huang Huang, Lewis Blake, Matthias Katzfuss

Earth-observing satellite instruments obtain a massive number of observations every day. For example, tens of millions of sea surface temperature (SST) observations on a global scale are collected daily by the Moderate Resolution Imaging Spectroradiometer (MODIS) instrument. Despite their size, such datasets are incomplete and noisy, necessitating spatial statistical inference to obtain complete, high-resolution fields with quantified uncertainties. Such inference is challenging due to the high computational cost, the nonstationary behavior of environmental processes on a global scale, and land barriers affecting the dependence of SST. We present a multi-resolution approximation (M-RA) of a Gaussian process (GP) whose nonstationary, global covariance function is obtained using local fits. The M-RA requires domain partitioning, which can be set up application-specifically. In the SST case, we partition the domain purposefully to account for and weaken dependence across land barriers. Our M-RA implementation is tailored to distributed-memory computation in high-performance-computing environments. We analyze a MODIS SST dataset consisting of more than 43 million observations, to our knowledge the largest dataset ever analyzed using a probabilistic GP model. We show that our nonstationary model based on local fits provides substantially improved predictive performance relative to a stationary approach.

Sunday 28.08.2022 09:00 - 11:05 Parallel Session H – SDS2022

SO004 Room Aula 4 STATISTICAL DATA SCIENCE (VIRTUAL) **Chair: Maria Brigida Ferraro**

S0189: **Change point inference for high-dimensional correlation matrix**
Presenter: **Zhaoyuan Li**, The Chinese University of Hong Kong, Shenzhen, China

The focus is on the problem of detecting and estimating a change point in the correlation matrix in a sequence of high dimensional vectors, where the dimension is substantially large compared to the sample size. We first propose a simulation-based approach to detect whether a change point exists or not. When we have witnessed a change point in the first step, a two-stage method is proposed to estimate the location of the change point. The first step consists of a reduction of the dimension to identify elements of the correlation matrices corresponding to significant changes, where the threshold is generated by a simulation-based procedure. In the second step, we use the components after dimension reduction to determine the position of the change point. This method can efficiently estimate not only the change point located in the middle of the sequence, but also that located at the tails, which will be very useful for online detection. Theoretical properties are developed for both approaches, and numerical studies are conducted to support the new methodology.

S0194: **Estimating specification parameters while learning unknowns, in a misspecified model**
Presenter: **Dalia Chakrabarty**, Brunel University London, United Kingdom
Co-authors: Niharika Paul

A parametric relationship between an output and an input variable is often motivated, while arbitrarily assigning values to those model parameters that cannot be learnt/estimated using the available data. We refer to such parameters as Specification Parameters, SP, and these are typically parametrisations of symmetries in the behaviour/structure of systems, where said symmetries cannot be informed upon by available observations, though the assignment of incorrect values - in the data - to SPs, affects output prediction. Existing literature permits testing for misspecification of a model, but there is no suggestion on how to learn/estimate values of such SPs. We present a new method for optimising SPs, while learning the other unknown parameters of the model using MCMC-based inference, given the data. At each chosen value of the SP, we compute a new divergence measure, between results obtained from learning given the empirical data, and that given the generated data, where the latter data is sampled from the probability density function of the observable, given the learnable parameters that are learnt using the empirical data. Divergence minimisation yields optimal SP values. We will illustrate our method by estimating the observationally-elusive parametrisation of anisotropy of the phase space of a real galaxy (NGC4494), while learning the distribution of mass of the galactic gravitational matter, thus resolving lack of identifiability between galactic mass and anisotropy.

S0169: **Identification of antigen specificity in single cell RNAseq experiments using biclustering methods for binary data**
Presenter: **Mohamad Zafer Merhi**, Hasselt University, Belgium
Co-authors: Dan Lin, Ahmed Essaghir, Ziv Shkedy

The single-cell RNA-sequencing technology allows the assessment of heterogeneous cell-specific changes and their biological characteristics. In our current study, we focus on single cell omics data for immune profiling purposes. T-cells exhibit unique behavior referred to as cross-reactivity; the ability of T-Cells to recognize two or more peptide-MHC complexes by the TCR. A CD8+ T Cell is defined as specific for an antigen if the cell binds to the antigen. The work is applied to single-cell RNA-seq data (publicly available in https://support.10xgenomics.com/single-cell-vdj/datasets/) consisting of CD8+ T Cells obtained using a state-of-the-art single-cell omics technology from 10X Genomics and our aim is to assess and understand the heterogeneous characteristics and the binding specificities of these CD8+ T Cells, i.e., we aim to identify the specificity of the CD8+ T cells to one (or more) antigen(s). For the identification of specific CD8+ T Cells, we proposed an unsupervised data analysis pipeline. Biclustering methods are applied to recover and explore the cross-reactive behaviour of T Cells and to identify a subset of cells which are specific to a subset of antigens. Clustering methods are used to link these subsets to the RNA-seq data. Furthermore, we discuss the challenges of the application and evaluation of clustering algorithms on the single cell RNA-seq data.

S0188: **Machine learning based mass imputation approaches for combining probability sample and nonprobability sample**
Presenter: **Sixia Chen**, University of Oklahoma, United States

Although probability samples have been regarded as the gold standard to collect information for population-based studies, non-probability samples have been used frequently in practice due to their low cost, convenience, and the difficulties in creating the sampling frames. Naive estimates based on non-probability samples without any adjustments may be misleading due to the selection bias. Recently, a valid data integration approach including mass imputation, propensity score weighting, and calibration has been used to improve the representativeness of non-probability samples. However, the effectiveness of mass imputation approaches depends on the underlying model assumption. We propose and compare several modern machine learning-based mass imputation approaches including generalized additive modeling, regression tree, random forest, XG-boosting, Support vector machine, and deep learning. Machine learning-based approaches have been shown to be more robust compared with the parametric mass imputation approach against the failure of underlying model assumptions. In addition, deep learning has been shown to be the most effective for handling hierarchical non-linear data structures. We evaluate our proposed methods by using both simulation study and real application.

S0158: **Effect of bootstrapping in sparse biological data and model selection via likelihood ratio test**
Presenter: **Mehmet Ali Kaygusuz**, Middle East Technical University, Turkey
Co-authors: Vilda Purutcuoglu

The likelihood ratio test (LRT) is very useful to compare with various models since it assesses the goodness of fit of competitive models based on the ratio of their likelihoods. From previous studies, it has been shown that LRT is preferable while alternative models are nested to each other and from recent studies, it has been observed that it is more computationally efficient than other model selection criteria. On the other hand, although the reduction in the description of the full model is challenging in complex models, it can be computational efficient when the difference between the number of parameters p and the number of observations n is large. We propose different bootstrap schemes to fill the underlying difference between n and p, and represent the data via a Gaussian graphical model. In model selection, we perform LRT due to its advantages in computational efficiency and high accuracy. As alternative models, we generate nested models from the results of k-means clustering so that we can evaluate comprehensively the combined effect of bootstrapping and LRT in the analyses of high-dimensional biological network data. In our assessment, we use simulated networks under different sparsity and sample sizes

SC014 Room Aula 3 STATISTICAL DATA SCIENCE **Chair: Ori Davidov**

S0178: **Semiparametric estimation for time series: A frequency domain approach based on optimal transportation theory**
Presenter: **Manon Felix**, University of Geneva, Switzerland
Co-authors: Davide La Vecchia

A novel approach is proposed for estimation in stationary ARMA processes. Our estimation is semi-parametric: we have a Euclidean parameter, but we do not assume any distribution for the innovation term. Working with the frequency domain approach, we use the Wasserstein distance (1-Wasserstein distance and 2-Wasserstein distance) to derive minimum distance estimators. To do this, we rely on the fact that the standardized periodogram ordinates are asymptotically independent and have an exponential distribution with rate one. We give heuristic arguments for their asymptotics and provide algorithms for their implementation. Monte-Carlo simulations illustrate the performance of our estimators, under different data generating mechanisms (e.g. leptokurtic underlying distributions, time domain additive outliers and frequency domain outliers). The numerical exercises highlight the improvements of our new estimators on the routinely-applied Whittles estimator.

S0179: **Portfolio choice when stock returns may disappoint: An empirical analysis based on L-moments**
Presenter: **Loriano Mancini**, USI Lugano, Swiss Finance Institute, Switzerland

The aim is to empirically examine the equity portfolio choice of investors with generalized disappointment aversion (GDA) preferences. Opposite to expected utility investors, GDA investors suffer large utility losses from suboptimal trading strategies such as equally weighted portfolios. These losses arise from the sensitivity to disappointing returns rather than from the threshold below which returns are perceived to be disappointing. Using a newly developed estimation method based on L-moments, we find that high order L-moments up to order ten, unlike conventional moments, have a substantial and economically sensible impact on portfolio choice.

S0193: **Mixtures, heavy tails, asymmetry and conditional heteroskedasticity in financial returns**
Presenter: **Fariborz Setoudehtazangi**, University of Padova, Iran
Co-authors: Massimiliano Caporin

Statistical modeling and analysis based on finite mixtures of symmetric distributions, especially normal mixtures, have been applied in many applications. In recent years, finite mixtures of asymmetric distributions have been developed as a powerful substitute for normal mixtures in a wide range of applications. The benefit of finite mixture models is to accommodate different characteristics, such as multimodality, skewness, kurtosis and heavy tails. Modeling financial data is considered a complex task, so not only working with financial returns is often involved in such a complex relationship with prior observations, but also the innovations, after fitting an appropriate model, drastically demonstrate skewness, kurtosis, heavy tail and multimodality. Financial returns usually reverberate a structure which can be logically illustrated with conditional heteroskedastic models, such as the Generalized Autoregressive Conditional Heteroskedastic (GARCH) process of Bollerslev. We propose a model adopting a mixture of asymmetric distribution for a financial time series characterized by conditional heteroskedasticity. The stochastic representation of the proposed model enables us to easily implement an EM-type algorithm to estimate the unknown parameters of the model. A comprehensive simulation study and real data sets are then conducted to evaluate the higher performance of the proposed method.

S0199: **Business cycle synchronization in the EU: A regional-sectoral look through soft clustering and wavelet decomposition**
Presenter: **Saulius Jokubaitis**, Vilnius University, Lithuania
Co-authors: Dmitrij Celov

The focus is on the sectoral-regional view of the business cycle synchronization in the EU – a necessary condition for the optimal currency area. We argue that complete and tidy clustering of the data improves the decision maker's understanding of the business cycle and, by extension, the quality of economic decisions. We define the business cycles by applying a wavelet approach to drift-adjusted gross value added data spanning over 2000Q1 to 2021Q2. For the application of the synchronization analysis, we propose the novel soft-clustering approach, which adjusts hierarchical clustering in several aspects. First, the method relies on synchronicity dissimilarity measures, noting that, for time series data, the feature space is the set of all points in time. Then, the "soft" part of the approach strengthens the synchronization signal by using silhouette measures. Finally, we add a probabilistic sparsity algorithm to drop out the most asynchronous "noisy" data improving the silhouette scores of the most and less synchronous groups. The method, hence, splits the sectoral-regional data into three groups: the synchronous group that shapes the EU business cycle; the less synchronous group that may hint at cycle forecasting relevant information; the asynchronous group that may help investors to diversify through-the-cycle risks of the investment portfolios. The results support the core-periphery hypothesis.

S0165: **Graphical linear models for paired comparison data**
Presenter: **Ori Davidov**, University of Haifa, Israel

Graphical linear models for paired comparison data will be discussed from both the frequentist and Bayesian perspectives. Estimation methods and large sample theory are described. Specifically, methods for quantifying the uncertainty in ranking procedures are emphasized. The methodology is illustrated using simulated data and applied to two data sets: a network metaanalysis example and to the ranking of teams in the National Basketball Association (NBA).

Authors Index